Indigenous, Modern and Postcolonial Relations to Nature

Indigenous, Modern and Postcolonial Relations to Nature contributes to the young field of intercultural philosophy by introducing the perspective of critical and postcolonial thinkers who have focused on systematic racism, power relations and the intersection of cultural identity and political struggle.

Angela Roothaan discusses how initiatives to tackle environmental problems cross-nationally are often challenged by economic growth processes in postcolonial nations and further complicated by fights for land rights and self-determination of indigenous peoples. For these peoples, survival requires countering the scramble for resources and clashing with environmental organizations that aim to bring their lands under their own control. The author explores the epistemological and ontological clashes behind these problems. This volume brings more awareness of what structurally obstructs open exchange in philosophy worldwide, and shows that with respect to nature, we should first negotiate what the environment is to us humans, beyond cultural differences. It demonstrates how a globalizing philosophical discourse can fully include epistemological claims of spirit ontologies, while critically investigating the exclusive claim to knowledge of modern science and philosophy.

This book will be of great interest to students and scholars of environmental philosophy, cultural anthropology, intercultural philosophy and postcolonial and critical theory.

Angela Roothaan is Assistant Professor of Philosophy at the Faculty of Humanities at Vrije Universiteit Amsterdam, the Netherlands. Her research focuses on African intercultural philosophy, philosophy of values and spirit ontologies. She has published books (in Dutch) on Spinoza, on nature in ethics, on truth, on spirituality and on ghosts/spirits in modern culture. Together with P. Nullens and S. van den Heuvel she published the edited volume *Theological Ethics and Moral Value Phenomena: The Experience of Values* (Routledge, 2017). Her philosophical blog can be found here: http://angela-roothaan.wordpress.com.

Routledge Environmental Humanities

Series editors: Scott Slovic
University of Idaho, USA
Joni Adamson
Arizona State University, USA
and
Yuki Masami
Kanazawa University, Japan

A full list of titles in this series is available at: www.routledge.com/series/REH

The *Routledge Environmental Humanities* series is an original and inspiring venture recognizing that today's world agricultural and water crises, ocean pollution and resource depletion, global warming from greenhouse gases, urban sprawl, overpopulation, food insecurity and environmental justice are all *crises of culture*.

The reality of understanding and finding adaptive solutions to our present and future environmental challenges has shifted the epicenter of environmental studies away from an exclusively scientific and technological framework to one that depends on the human-focused disciplines and ideas of the humanities and allied social sciences.

We thus welcome book proposals from all humanities and social sciences disciplines for an inclusive and interdisciplinary series. We favour manuscripts aimed at an international readership and written in a lively and accessible style. The readership comprises scholars and students from the humanities and social sciences and thoughtful readers concerned about the human dimensions of environmental change.

Indigenous, Modern and Postcolonial Relations to Nature

Negotiating the Environment

Angela Roothaan

First published 2019
by Routledge
2 Park Square, Milton Park, Abingdon, Oxon OX14 4RN

and by Routledge
605 Third Avenue, New York, NY 10017

First issued in paperback 2020

Routledge is an imprint of the Taylor & Francis Group, an informa business

British Library Cataloguing-in-Publication Data
A catalogue record for this book is available from the British Library

Library of Congress Cataloging-in-Publication Data
A catalog record has been requested for this book

Typeset in Garamond
by Wearset Ltd, Boldon, Tyne and Wear

ISBN 13: 978-0-367-72849-6 (pbk)
ISBN 13: 978-1-138-33777-0 (hbk)

Contents

Preface and acknowledgments

Preface

This book offers a philosophical analysis of obstacles and opportunities for dialogue concerning the environment in our present globalizing world. Initiatives to tackle environmental problems cross-nationally are often challenged by economic growth processes in postcolonial nations, further complicated by fights for land rights and self-determination of indigenous peoples. In order to survive, these peoples try to counter the scramble for resources which threatens their environment. Sometimes they also clash with environmental organizations that aim to bring their lands under external control, e.g., to protect rare animal species. Consequently, disorderly and often violent confrontations take place in remote corners of the earth. Remote in the sense of: far away from the media and the powers that are centered in global cities. Connecting more and more through new ways of communication these peoples increasingly speak out, not only for their own survival, but also because they see that 'we,' those representing modern, global culture, destroy our common habitat by destroying theirs.

At the background of these confrontations and destruction are different ways to relate to nature; that is where this book comes in. It is high time to develop an intercultural environmental philosophy that focuses on possibilities to negotiate this environment that is ours, communally, as humans. We need to analyze and critique the epistemological claims entailed by the different positions, as well as their relation to political-economic interests. Such a philosophy may clarify the possibility conditions for negotiations between the parties involved. One party claims that all beings are a meaningful part of the web of energetic relations we call nature, which means they all are needed to be able to enjoy life on this planet. The other party, claiming that the human race stands apart from and over against nature, a nature perceived as hostile and to be subjugated and used for resources at will, has power on its hands. It is their position that is seen as the root cause of our present environmental problems, not only by modern critics, but very explicitly by representatives of indigenous peoples, such as shaman Davi

Kopenawa,[1] who claims even those who live in modern societies need protection and cure from someone like him:

> It is true. The shamans do not only repel the dangerous things to protect the inhabitants of the forest. They also work to protect the white people who live under the same sky. This is why if all those who know how to call the *xapiri* die, the white people will remain alone and helpless on their ravaged land, assailed by a multitude of evil beings they don't know.
>
> (Kopenawa and Albert 2013 [2010]: p. 404)

It is time to find ways to include the voices of thinkers and activists like Kopenawa in environmental philosophy.

To get to listen to the voices of shamans, and to do justice to shamanistic[2] ways to be in the world is not an easy matter for the moderns.[3] The power systems of modernity, including its politics of epistemology, are too much entangled in all aspects of our lives, and in too complex ways, to make listening directly possible. To get to our aim, we need to diligently and intelligently disentangle where and how these systems dictate the views, the practices and the material structures in which modern life is lived. First and foremost, we need to clarify the different relations humans have to nature – modern, science-based ones, as well as shamanistic or spirited ones. To make things even more complex, both ways to relate to nature at times intersect with institutionalized religions, who have their own views of nature in the framework of a supposed divine order. To get to such a clarification, as well as to the critique of modern relations to nature, and to the laying out the possibility conditions for negotiating the environment in which we live, we need to carefully analyze those concepts that may reproduce existing power relations. Among such concepts are notably the ones we have used here freely: 'indigenous,' 'shamanic,' 'spirits,' 'postcolonial,' 'modern,' 'animist,' which all may sustain the othering of those who are threatened by modern cultures and their economic gospel of material progress. I will get to such clarifications where they are needed in the course of this text, but most can be found in the final section of Chapter 1. One of the central concepts, that of 'nature' I will leave undefined. This concept has translated many different meanings throughout the history of philosophy (Roothaan 2005). By leaving it undefined, an open space is created for the intercultural dialogue this book is working toward. In this preface I will confine myself to pointing out the central issues at stake and sketch the main outline of the book.

The most well-known of the many struggles over the environment may well be 'Standing Rock,' named after the protesters' camp built to halt the Dakota Access Pipeline. Native Americans from different peoples joined environmentalists and Lakota people who lived on the land through which consecutive US governments were planning to build a new giant pipeline to transport oil from the North (Canada) to the South.[4] The camp, set up in 2016, remained in place many months, until the protesters were removed by the forces of order. Many similar struggles are taking place that remain outside of the attention of the media, and research indicates that environmental defenders are killed almost daily all over the world.[5] Defenders often are representatives or members of indigenous peoples, whose lands and ways of life are under threat, for example we may think of the protests against the Belo Monte Dam in Brazil, struggles of Aboriginal peoples in Australia (e.g., the one that resulted in the return of land rights to the Martu people), conflicts concerning the environment in parts of Russia (e.g., the Buryat people), Africa (e.g., the San, or the Baka people) and Europe (the Saami people). For these peoples it is not only about survival, they also protest the modern way to treat nature as a source of resources and commodities and consequently of wealth in terms of money. They hold humans and animals (actually, all natural phenomena) to be interconnected physically and spiritually. By investigating indigenous relations to nature, as they express themselves in shamanistic or spirit ontologies, the struggle between different ways to live on this earth can be taken to the level of a politics of epistemology, which in its turn will uncover the intrinsically political nature of Western philosophy.

Such an investigation should make use of a dialogical and deconstructive hermeneutics, as it has been developed in the young field of intercultural philosophy, but also of the analysis of power relations as propounded by postcolonial and critical theory. Without these approaches it will not be possible to secure the preconditions for negotiation. This book proposes to develop an environmental philosophy that offers the groundwork for such negotiation, meanwhile also discussing some actual cases where the positions clash. These are taken mainly from African contexts – not only because here we find the contrary positions actively and explicitly opposing each other today, but also because intercultural philosophical dialogue on human relations to nature was probably first initiated here, in the debates following the pre-decolonization publication of a work describing 'Bantu ontology' as a spirited ontology of vital force inherent in nature, and in humanity as prefiguring ontological investigations of spirit ontologies by present day anthropologists.

Epistemologically speaking, the core of the struggle over the environment in postcolonial contexts is whether shamanistic indigenous ontologies that hold non-human nature to be spirited or animated will be able to successfully negotiate their interests in discourses on environmental protection that are dominated by modern secular ontologies. From a modernist point of view, shamanistic or spirit ontologies are considered to 'believe' in the existence of a spirit world next to and interacting with the empirical world. From their own point of view, such a dualism doesn't arise in the same way, as the distinction between a material and a spiritual realm is not made similarly.

Instead we find in spirit ontologies varying combinations of distinctions between what is real/living/an agent and not real/dead/passive.

Such ontologies have been studied in cultural anthropology as the foundation of 'premodern' or 'non-Western' worldviews. One can contend, however, that they also represent suppressed realities of modern Western cultures, and that they should be included in a transcultural, existential understanding of humanity. Consequentially, the characterization of such ontologies as premodern or indigenous, defining them in time and place, will be challenged as well. Only then can we investigate how the epistemological claims of spirit ontologies can be made to bear in a globalized philosophical discourse. Thus we will critique the exclusive claim to knowledge of (materialist, empiricist) modern science and philosophy. Drawing from sources such as cultural anthropology, pluralist ontology, intercultural philosophy and postcolonial and critical theory I hope to arrive at this aim. Several authors from within the Western philosophical world have already provided critiques of the epistemological approach of technologically-oriented science – as colonial (Sandra Harding), as forgetful of our spiritual and material relations with the world (Bruno Latour) and as expressive of anonymous systems of power techniques (Michel Foucault). Even more important are the postcolonial critiques of racism in modern philosophy (Aimé Césaire, Frantz Fanon, Edward Saïd, Emmanuel Eze). The mentioned authors have proposed necessary alternatives to understanding humanity – to move beyond the colonization, submission, abuse and murder of those that modern cultures dehumanized and dehumanizes.

Next to these critical and constructive movements, the book will discuss recent attempts of cultural anthropologists to reconstruct the ontological deep structures of the worldviews and practices of shamanistic indigenous cultures (Eduardo Kohn, Eduardo Viveiros de Castro). Making use of Western philosophy, ranging from American pragmatism to French poststructuralism, they attempt to clarify ontologically the possibility that forests think, that humans communicate with animals as ontological relatives, and enter into contact with spirits of diverse origin. In so doing they repeat certain aspects of the work of Belgian missionary Placide Tempels, whose *Bantu Philosophy* (1959 [1946]) opened a 70-year-long debate on the necessity/possibility to culturalize philosophy in the African context. We should evaluate whether present attempts to culturalize philosophy, such as those of Kohn and Viveiros de Castro, bring new insights in comparison with Tempels's work, and/or ignore insights he gained in his research. Something to note is also the effect of their relative neglect of spirituality/religion in their understanding of indigenous ontologies, an aspect that Tempels centered while he discussed the political and economic context of colonialism and evangelization. Intercultural philosophy, being a historical, existential and situated practice, should always be critically aware of the political and economic frameworks in which intercultural encounters take place. Including spirits or excluding them is not just a matter of 'culture,' but also an expression of power relations, and ways to contest those relations. Therefore, intercultural philosophy will also always include a political critique.

This book begins by looking into the kind of world experience that allows for trance states, in which animal spirits and other spiritual agents may enter in communication with humans. I pose the question: what is at stake in spirit ontologies? Because I developed my work as a critical discussion with the 'Western' canon which provided the framework of my education, I will enter the space of shamanic experience through the autobiographical narratives of two writers who discovered it as an answer to the Western-styled culture in which they were educated as well, answering to what they knew to be beyond its boundaries: Robert Wolff and Malidoma Somé. The first chapter discusses how we can speak of the animated, the spirited, and how interdisciplinary researchers in the last two centuries have searched for ways to express what spirit ontologies have to say in philosophy. In the second chapter I will elaborate on two strands of work that aim to give a voice to spirit ontologies. I will discuss works that represent what is called the *ontological turn* in cultural anthropology, next to those authors who defend a new kind of *animism* as a form of environmental ethics. I will also relate these strands to the aim, developed in philosophical phenomenology, to overcome the dualisms of modern thought. These different approaches will be confronted with each other to see what they may add to the articulation of possibility conditions to begin to negotiate the environment.

In the third chapter we will turn to the early modern history of ideas (of the West), to understand how the demarcation of modern scientific knowledge and of Enlightenment ideals of personal formation went hand in hand with a move to ban or exorcize the spirits from Western culture. The major turning point in the banning of the spirits lies in *Dreams of a Spirit-Seer* (2002 [1766]), Kant's criticism of the Enlightenment visionary Emanuel Swedenborg, in which he denounced the epistemic claims of his contemporary who has written extensively on the spirit world he perceived beyond and behind the empirical world. As *Dreams of a Spirit-Seer* (2002 [1766]) preconfigured the main arguments of Kant's critical works, it bears testimony of how modern epistemology both manifests and flows from the banishment of a spirit ontology. All the same, Kant scholars are beginning to argue that Kant failed to expel Swedenborgian ideas completely, and even relied on them to construe his moral philosophy. Referring to the work by Peter Park and Emmanuel Eze I will also investigate the racist politics of epistemology that are at work in Kant's Enlightenment project and question how these relate to his account of spirit ontologies.

After 200 years several Western thinkers have been seen to make an attempt to re-introduce spirit terminology, being frustrated by the philosophical and scientific restrictions of the Kantian delimitation of valid knowledge. In Chapter 4 I will discuss three of them: William James, Carl-Gustav

Jung and Jacques Derrida, who took rather different routes to do so. The aim of this discussion is to show that (a) the spiritual will always return whenever it is conjured away from the realm of valid knowledge, and (b) that there are different possible approaches to opening up Western thought, in order to make such a return possible. The differences between the approaches – pragmatist ontology, psychology of the collective subconscious and deconstruction will be investigated for their respective benefits to the disclosure of culturally mediated experiences of and communications with animated beings, disputably called spirits, and their inclusion in an increasingly intercultural philosophy. Special attention should be given to what the different relations to the spiritual and to nature mean for our (human) relations to non-human-animals. It turns out that animality presents the imaginary boundary used by many Western philosophers to separate our embodied experiences from the purely rational mind, *and* to separate human from non-human living beings. In Chapter 5 I will discuss what a deconstructive approach as well as a decolonial approach bring to the question of animality. Although radically different in their 'methodology,'[6] the outcomes of both approaches come together where they problematize the overlap of racist and speciesist othering of the non-human-animal and the animalistic nature of human beings. Both are also needed to blur the dividing lines that kept spiritual encounters between human and animal out of the realm of valid knowledge, and to criticize modernist anthropology as 'the study of man.'

In Chapter 6 we turn to Placide Tempels and his 1946 work *Bantu Philosophy* (1959), which is arguably a precursor of what is now called the ontological turn in anthropology. In this book Tempels reconstructed the spirited world experience of those he came to convert with the help of Western philosophical concepts. His religious aim to not dehumanize his African interlocutors as this would make their true conversion impossible, made him cross standing delineations of theology and philosophy, as well as change both disciplines from within and Africanize them. Tempels also crossed the disciplinary boundaries between ethnology and philosophy, decades before cultural anthropologists commenced doing the same. Rereading Tempels will add meaningfully to the question: can philosophy be critical *and* embrace spirit ontologies? In the seventh chapter we move on to discuss a present day, real-life case of spiritual communications between humans and non-human-animals, which situates itself in the framework of postcolonial political, legal and economic conditions. The case concerns southern African elephants who featured in the life story of conservationist Lawrence Anthony, who got in close contact with a herd he invited to his reserve. When he died, the elephants walked for hours to reach his house, apparently to mourn him. The question their behavior raises is: how to speak of elephants – and how to speak with them – in the conditions of fenced in lands, postcolonial legal and political structures, and decolonizing and conservationist movements? What are the possibilities to enter into spirited communication today with animals such as elephants? From the case of elephants, we move to the case of trees in West Africa, in Chapter 8, where I will argue that Christian, Islamic and traditionalist approaches of nature *and* the secular discourse that reigns over conservationist policies, must negotiate the environment if they want to have a positive effect on our relations to nature. Through the discussion of this case, we will find that similar political, economic and legal frameworks that reign over the understanding of human–animal relations recur in the case of human-vegetal ones. When, through deconstructive and decolonial approaches, modern boundaries fencing off the animal and the spirited become blurred, so do the boundaries that exclude living beings from the category of 'animals' – trees and other plants. Arguably, Christianity and Islam do not only suppress, as is often thought, but also mediate shamanic experiences of the world, and of the relations between humans, animals, plants and the spiritual and the spirited.

After having looked into these different cases where modern and indigenous relations to nature collide in postcolonial contexts, it is time to draw the outlines of a framework that enables negotiating the environment. Although I started out from intercultural philosophy which favors *dialogue*, I have come to stress the need for *negotiation*. This need stems from the warlike situation in which modern cultures find themselves with indigenous defenders of natural environments, as well as with nature itself. To arrive at dialogue, negotiations are in place first. The framework needed to hold them will be drawn up in the concluding ninth chapter, and will consist in a postdeconstructivist anthropology of anim(al)istic life. There we can draw the consequences of the deconstructive and decolonizing movements made in the preceding chapters. Whereas Kant brought the question of reality down to that of rational humanity, our work to blur the boundaries between animals, humans and spirits will now have changed anthropology – and, in its wake, what counts as real. It will be shown that countering the othering of spirit ontologies and questioning the definition of the human-animal are complementary actions. Both are needed, in concert, to bring awareness of what structurally hinders a truly open intercultural exchange in philosophy worldwide, and to prepare the field where people with different epistemologies and ontologies that cut through cultural identities can meet to discuss our shared and most pressing questions concerning the environment. A redefined intercultural and postcolonial philosophy will finally be able to reach beyond academia to real-life 'indigenous' struggles for land rights and for cultural survival, struggles that are motivated by the hopes of indigenous peoples to survive the grasp of the scramble over resources, by hopes for some kind of repair and retribution for the genocides and removals (over the past centuries) of first nations peoples from their lands, but that are also inspired by their hope to convince the moderns to review their relations to nature, in order to prevent destruction of the earth as a viable place for all and any forms of earthly life.

Acknowledgments

To acknowledge all people who helped, inspired and sustained me while I was in the process of research and writing this book is an impossible task. For this reason I will not try to name each and every one of them, and express here in general: thank you, all of you, who are in my life. What I will try to mark here is the chain of encounters that most directly led to the realization of this book and influenced the form it has taken. When, in 2013, I formed the plan to rework my Dutch book on spirits in modern Western culture (Roothaan 2011) into an English book that should focus on spirit ontologies and make use of the literature in anthropology and religious studies I had been reading since then, I started meeting people who played a significant role in how this project took shape. First, at a conference in the Netherlands, I met Gabriel Fonteles, a historian from Brazil working on indigenous Amazonian epistemologies, whose reading suggestions added focus to my work. Around the same time I encountered David Ludwig as a new colleague at work, whose knowledgeability of debates on indigenous ontologies from a philosophy of science perspective inspired me and helped me along.

Things gained momentum when I got acquainted with Michael Eze, then in the USA, who introduced me to the real world of African philosophy, suggesting me to participate in a conference in Calabar, Nigeria, where I presented the earlier version of what has become Chapter 8, the chapter on trees. When its first version was published (in Chimakonam 2017), Leila Walker, from Routledge's *Environmental Philosophy Series*, invited me to submit a book proposal. Rewriting the spirit ontologies-plan to fit the series's focus added decisively to the work. At the Calabar conference I had also met Pius Mosima from Cameroon, with whom I developed a collaborative research project on Tempels's *Bantu Philosophy* (1959 [1946]). The work on this project has benefited Chapter 6 most directly. Finally I met with and entered into several editing and organizing projects in African and intercultural philosophy with Louise Müller, working at Leiden University, which provided a stimulating context for this work.

In the process of submitting my proposal Rebecca Brennan took over as editor, and with the assistance of Julia Pollacco, she coached me through the work of planning, editing and other material aspects of getting the work done. I am grateful for the opportunity to publish my book in this series, and for the professional guidance and support. I also want to thank my colleague Govert Buijs, who gave helpful comments in an earlier stage to improve the book proposal, and my colleagues Peter Versteeg and Hans Radder for being always particularly supportive of my research work. Last but not least I want to mention and thank the board of editors who approved of my proposal, and the three anonymous reviewers who made my day when I read their deeply informed comments and positive criticisms, that – while writing – helped me to keep focused on the 'little' things that actually are decisive to create precise and informed work: to continuously aim at a consistent clarification of concepts used, and to remain aware of the many interrelated interdisciplinary discussions in which this book aims to intervene and to which it hopes to add.

Chapter 5 was adapted from Roothaan, A. (2017). 'Aren't We Animals? Deconstructing or Decolonizing the Human–Animal Divide' in M. Fuller, D. Evers, A. Runehov and K-W. Saether (eds.) *Are We Special? Human Uniqueness in Science and Theology*. Cham: Springer, pp. 209–20.

Chapter 7 was adapted from Roothaan, A. (2018). 'Hermeneutics of Trees in an African Context: Enriching the Understanding of the Environment "for the Common Heritage of Humankind"' in J. O. Chimakonam (ed.) *African Philosophy and Environmental Conservation*. London and New York: Routledge, pp. 135–48.

Chapter 8 was adapted from Roothaan, A. (2019a). 'Decolonizing Human–Animal Relations in an African Context' in M. Chemhuru (ed.) *African Environmental Ethics: A Critical Reader*. Springer, (forthcoming).

A note on language used

As the non-inclusive language of modern Western thought is one of the things contested in this book, I decided not to erase it when discussing it – so I wrote 'man' and 'mankind' whenever discussing modernist thinking on the human-animal and its relations to nature. I used humankind where I am closer to an alternative approach, but have attempted to use even more diverse wording for who is speaking, whenever it was possible to open thinking spaces for epistemic interculturality. In citing I follow the spelling of the works cited, which may differ from the spelling adopted throughout the main text of this work.

Notes

1 Davi Kopenawa is a Yanomami shaman and rights defender of his people from northern Brazil. French Anthropologist Bruce Albert, who has known him over a period of decades, has published with footnotes the 'auto-ethnography' of Kopenawa, on the basis of hours of recorded conversations.
2 I will use the word 'shamanistic' when I speak of cultures, peoples and practices in a more abstract sense. When discussing specific rituals, practices, insights as carried out or begotten by shamans, I will call them 'shamanic.'
3 The signifier 'the moderns' is taken from Latour's 2013 study of modern culture by an imaginary anthropologist, *An Inquiry into Modes of Existence*. I will use it throughout this book to indicate that those living in modern cultures can be seen as a globalizing 'tribe.'
4 In her book *Standing Rock*, journalist Ekberzade has documented the protests at Standing Rock within the historical framework of native American resistance to their colonization. Cf. Ekberzade 2018.

5 Cf. the 2015 report *How Many More?*, published by Global Witness, that has a number of 116 environmental activists killed in 2014 alone, of whom 40 percent were indigenous. Hydropower, mining and agri-business are the most common cause of conflict. Following their latest report, almost the double number of 'earth defenders,' 207, were recorded to have been killed worldwide.

6 Deconstruction opposes the idea of methodology, that is, the idea that a set of rules how to go about knowing something may ever be given apart from or 'before' the knowing itself. Decoloniality, stressing the very political nature of all knowledge, does not deny the idea of methodology, but treats it as situated in a stance in and to power.

Chapter 1

A world of motion and emergence

An outline of what's at stake

Border crossings

Philosophy does not start with wonder. It starts with being touched. Of this I was reminded when I read, a few years ago, the first page of the first chapter of *Where Spirits Ride the Wind* (Goodman 1990). In this book, Felicitas Goodman, an out-of-the-box anthropologist, describes the shock she experienced upon finding out, at age 12, what becoming adult meant: 'The magic time is over.' I recognized what she described:

> ...all of a sudden and without the slightest warning, I realized that I could no longer effortlessly call up what in my terms was magic: that change in me that was so deliciously exciting and as if I were opening a door, imparting a special hue to whatever I chose.
>
> (Goodman 1990: p. 4)

The experience of 'magic,' of a world shining and bristling with life, is – it must be clear – not just an aesthetic mode of being. It is not the 'receiving,' like in a revelation, of a spiritual reality – imperceptible for the uninitiated. It springs rather, as Goodman's description makes clear, of her rediscovery of the 'magic time,' which means participating in a mutual, playful, dialogical relationship with beings around us – opposed to the 'normal' modern relationship to nature, which filters out conscious and intentional agency of any being but the rational ones, making the world around us, even though the beings in it move and interact, in a sense inert and alien. This filtering-out is characteristic for a secularized technological interference with nature. It presupposes (for example) that the tree I want to fell to build a shelter will not strike back at me. In the non-secularized, 'magical' experience, I will need to address the spirit of the tree, to conjure it or atone it to fence off retaliation. Such address can take varying forms, like making offerings (giving something back for what I take) to gods and deities, saying ritual prayers, or warding off adversities by placing a power object at a place where danger may most likely enter our world.[1]

For Goodman the loss of the ability to enter a world '...where I could always feel the comfort of invisible presences around' (1990: p. 5) came with a shock, and her return to that openness for the spiritual world was marked by unexpected events. After she had moved from Europe to the US, and was visiting a place in New Mexico, she felt invited by presences living there to buy the land and build a house on it. From that time on she increasingly became aware of how animals and spirits gave her messages to heed, which set her on a path of rediscovery of ways to learn about life and the world through shamanic trance states. With her students she investigated how body postures portrayed by ancient statuettes from horticultural and hunter-gatherer societies lead to varying types of trance, which offer insights and help with things like healing, knowledge, protection, attack, etc. Thus she rediscovered what we will call in this book a *shamanic way to be in the world*.

The results of Goodman's uncommon empirical research were presented in the above-mentioned book, where she characterizes the world she came to live in as a world of motion and emergence, contrasting it with the passive, motionless *world of being* that had been sought after by philosophy and science ever since the ancient Greeks. Goodman's observations do not stand alone – one can find a similar turn away from the reductive aspects in modernity in the posthumous book by Paul Feyerabend, *Conquest of Abundance* (1999). About the ontology that has for the larger part characterized what we call modernity, he writes:

> The search for reality that accompanied the growth of Western civilization played an important role in the process of simplifying the world ... this search has also a strong negative component. It does not accept the phenomena as they are, it changes them, either in thought (abstraction) or by actively interfering with them (experiment).... In both cases, things are being taken away or 'blocked off' from the totality that surrounds us. Interestingly the remains are called 'real,' which means they are regarded as more important than the totality itself.
>
> (Feyerabend 1999: p. 5)

In contrast with this way to understand and manipulate being, Feyerabend aims to get a wider reality into view, which includes ways of being in the world that have been cut off by modernist orthodoxy as primitive. In an attempt to evoke the experience of this reality, he sums up:

> The world we inhabit is abundant beyond our wildest imagination. There are trees, dreams, sunrises; there are thunderstorms, shadows, rivers; there are wars, flea bites, love affairs; there are the lives of people, Gods, entire galaxies.... Only a tiny fraction of this abundance affects our minds.
>
> (Feyerabend 1999: p. 3)

Feyerabend is very much aware, and I follow him in this, that opening up to the wider reality, which means opening up to shamanistic or spirit ontologies, presupposes bracketing the exclusionary principles that characterize modernism. This again means that we undefine the boundaries of modern culture, opening up the possibility that inside modern culture itself the phenomena in all their richness are still perceived, by some, at certain moments, and also expressed in texts and art forms which the orthodoxy classifies as 'esoteric.' In line with this an important point in this study will be that although modernity aimed to ban spirited beings from reality, it has never completely succeeded in doing so. Precisely this may provide a point of connection that will make the hoped-for dialogue, and the negotiations that have to precede dialogue, possible.

Another interesting recent attempt to describe and critique the limits of modernity is found in the works of Bruno Latour, especially in his *An Inquiry into Modes of Existence* (2013). In this quasi-anthropological study, its author investigates the implicit presuppositions, the deep values of the tribe he calls 'the moderns.' I will make use of this name here, as it provides us with the ability to speak about a certain life world, with its ontologies, its ways of life, its beliefs and practices, without having to essentialize a *worldview*, the 'modern' worldview, in the same abstracted manner that modern abstraction seduces us to. Latour's work is, like this book, equally motivated by a concern about the environment, and it also looks for a way to repair the 'othering' of indigenous ways of being in the world, as this seems to be a point at which mending is necessary. My approach differs in many respects from Latour's as well, most importantly I challenge his contention that the moderns can/may enter, without a preceding analysis of the violent nature of their conquest of the world, into negotiations, and even without reparations. His contention follows from the fact that he remains on the 'clean' playing field of philosophical reflection. After his investigation into the deep values of the moderns, he claims that

> ...only then, might we turn back toward 'the others' – the former 'others!' – to begin negotiation about which values to institute, to maintain, perhaps to share ... Together, we could perhaps better prepare ourselves to confront the emergence of the global, of the Globe, without denying any aspect of our history.
>
> (Latour 2013: p. xxvii)

Latour is moving too fast here. As long as forests are being cut at a terrifying rate and the indigenous peoples who live in them still have to struggle to get legal recognition of their autonomy and ownership, the moderns offering them to negotiate common values is precocious. Indigenous activists for the environment are being killed for their work all over the world, which makes the situation between the parties involved to have more characteristics of a war than of a clash of values, and to end wars negotiations are in place before dialogue can be done.

To prepare for dialogue, it is therefore of utmost importance to include in our analysis the systems of power, the politics, the economic realities, the ideologies and instruments and people that have not yet ended this war. This is what the present postcolonial situation asks of us. With the word postcolonial I do not mean that colonialism is in the past and we can now make a fresh start among peoples. The eruption of violence, of domination of people and of nature by 'the moderns' that colonialism was, has not only left its marks on the actual lives and the minds of those who inherit its effects, but also has put in place so many legal structures, value systems, moralities, trading practices, migration patterns, etc., that are still effectively dominating our relations, even after institutional colonialism has been abolished. I will touch on such structures when I describe specific cases (trees, elephants) in which the environment is negotiated. I urge philosophy to move beyond an analysis of values or ideas only, and to also take the contexts into account which provide the conditions of success or failure of the negotiations that are to be held. Decolonization is still in process, and we should still heed, continuously, a warning like that of Fanon, that

> In its narcissistic monologue the colonialist bourgeoisie, by way of its academics, had implanted in the minds of the colonized that the essential values – meaning Western values – remain eternal despite all errors attributable to man.
>
> (Fanon 2004 [1961]: p. 11)

We have to remain aware of the effects of colonial knowledge systems that, in a two-tier move (by means of modern epistemology and of ethnology), have denied spontaneity and conscious action to non-white[2] people, and to non-human nature – identifying spirited ontologies to be the primitive worldview of the natives. The consequences of this colonization of minds and bodies for the peoples that were caught in this conceptual trap, have been well-analyzed by authors such as Fanon, Saïd and others. The consequences for non-human nature have, up till now, been analyzed well enough, but are almost absent in academic philosophical literature. This book wants to change this by undoing the two-tier trick, showing not only that 'we have never been modern' (Latour 2013), but also that there is a chorus of natural spiritual voices to be heard around us – voices that are still heard by those who practice shamanic ways to be in the world. This is what is at stake, to seek to hear the voices of indigenous peoples, and consequently of the spirited beings that they hear or used to hear. To be able to do this, we have to be ready to bracket, and negotiate the reach and power of modernist approaches to nature and to humanity.

In order to take shamanistic ontologies and those who hold them seriously as negotiating partners, modernity should stop to describe them from the outside, as 'other.' This means trying to take their voices seriously. To prepare for such a step it may be helpful to understand what is at stake existentially, for actual human beings, if they cross over to the 'indigenous' side. There are many stories of this kind of border crossings, and I will highlight two of them here: the first one of a Dutch psychologist who came to work for developmental organizations and over time became increasingly involved with indigenous people and their ways to relate to the world. Robert Wolff, who died in 2015 at the age of 94, having lived the final stage of his life in Hawaii, feeding himself on what he could find in nature, wrote down his 'conversion story' to an indigenous way of life in a collection of stories titled *Original Wisdom* (2001). The second one is of a Burkina Faso (then Opper Volta) – born shamanistic coach and writer, who saw himself destined to go through the Western education system, to finally leave the life of an academic in the USA in an attempt to bridge the worlds of knowledge. Malidoma Patrice Somé's autobiography *Of Water and the Spirit* (1994) became a classic among spiritual seekers. Nowadays, aged 61, he makes a living introducing Westerners to shamanic rituals. Presenting the self-narration of the crossover of these two men, and the healing knowledge they wanted to share with their audience, I will analyze the ways in which they crossed cultural, geographical and ontological boundaries, while they felt their destiny to lie in the rediscovery and/or active restoration of 'ancient' ways to be. What interests me here is how the powers of modern culture and those of the indigenous ones clash, and lead them into *unidentified territories of knowing*, to find another way to relate to nature, to all that is living and of which the individual human being is a part.

Tiger

Robert Wolff embarked on his journey to expand his knowledge of the predicament of today's world with the critical approach of a scientifically trained researcher, an approach he never lost, even after his close encounters with the Sng'oi people of Malaysia who changed his life. In the foreword of the report of Wolff's explorations of a wider reality, Thom Hartmann writes of those who treat spirituality as just another consumption article on the global market:

> It's even fashionable nowadays for First World eco-tourists to visit remote parts of the world, spend a week or two with an indigenous shaman, smoke a few plants, see a few hallucinations, then come back to declare themselves shamans and develop large followings. Shamanism for self-growth, shamanism for business, shamanism to build wealth and power – it's popping up all over…

> (Wolff 2001: p. ix)

Wolff understood his own journey to be different from such instant-shamanism. All the same, it also started in the trail of the Western-dominated political-economic expansion now named globalization. Wolff starts his story on the island of Sumatra, colonized by the Dutch, where his family lived as 'colons.' That he felt more warmth and belonging in the world of the local 'servants' than with his own kin is a story that doesn't stand on its own. Present day norms about parenting in the West, which normatively prescribe sharing and guiding ones children on a close by, affectionate and day to day basis, are far from the Victorian style treatment of children in wealthy families up till World War II. While mothers and fathers were busy pursuing the bourgeois life, young children were left with house-maids and other personnel, in the 'Old' world as well as in the colonies. In little Robert the exposure to a different way of being and understanding meant the first initiation to his life long search for a non-Western mode of living.

> ... it was impressed upon me that humans always exist within a larger context. I knew that people, despite great differences, are related as humans, as we are related to the animals and plants around us.
>
> (Wolff 2001: p. 1)

After working in several countries as a psychologist for several organizations, the great change in his life came when Wolff lived for several years in Malaysia. Driven by curiosity he made contact with the seminomadic Sng'oi people, who lived deep within the forest, hoping to be left alone by modern society. Being introduced by a town dweller who is related to them, the Sng'oi however receive Wolff friendly, and when they start to accept him as a regular visitor, he learns their language and commences to stay over for several days at a time whenever he is in their neighborhood. In his book he describes many aspects of Sng'oi knowledge and ways of life. Most interesting is his description of their collective sessions of dream interpretation in the morning, which tell them things that will happen or that are to be done. Wolff finds that they master ways of intuitive knowing that seem strange or unbelievable to modern peoples – like that day when all of them go on a long walk to see a giant forest flower that blooms only one day. Several of them had 'seen' the flower in their dream, so they knew the day it was blooming. Another story tells how always when Robert approaches the village on one of the jungle paths, someone is awaiting him before he reaches his aim. There are no ways to let them know beforehand that someone is coming, still they know when he does.

After some time Wolff is invited by Ahmeed, the shaman of one of the villages to be initiated in the shaman's way of knowing. The instruction consists of nothing but long days of walking in the forest with Ahmeed, without much talk and without food or drink. It takes several visits with several walks on which Robert doesn't notice any change, and even a period in which he distances himself from the process out of frustration. When he returns for a final try however, on the second day of walking suddenly a new way of experiencing opens up to him:

> I stopped abruptly. The jungle was suddenly dense with sounds, smells, little puffs of air here and there. I became aware of things I had largely ignored before. It was as if all this time I had been walking with dirty eyeglasses – and then someone washed them for me.... I could smell things I had no name for. I heard little sounds that could be anything at all. I saw a leaf shivering. I saw a line of insects crawling up a tree.
>
> (Wolff 2001: p. 156)

Then Ahmeed asks him whether he wants to drink, and points out that Robert now is able to sense where water is with his inner sense, which he then succeeds to do:

> As soon as I stopped thinking, planning, deciding, analyzing – using my mind, in short – I felt as if I was pushed in a certain direction. I walked a few steps and immediately saw a big leaf with perhaps half a cup of water in it.
>
> (Wolff 2001: p. 157)

What's more, he describes this impression not only as a practical bit of empirical information (there is water in the leaf), but instead as part of an encompassing metaphysical experience of the interconnectedness of all life:

> My perception opened further. I no longer saw water – what I felt with my whole being was a leaf-with-water-in-it, attached to a plant that grew in soil surrounded by uncounted other plants, all part of the same blanket of living things covering the soil, which was also part of a larger living skin around the earth. And nothing was separate; all was one, the same thing.... The all-ness was everywhere, and I was part of it.
>
> (Wolff 2001: p. 157)

Soon after that first opening up to the living forest around him, Wolff has another significant experience. As he had been 'away' a few minutes in his 'mindless' experience, he finds his companion has disappeared in the forest. His impulse to panic however stops before it develops fully, as the feeling of interconnectedness with all around him is so strong that he feels absolutely safe. He realizes not only that he will 'know' where the village is when he needs to go home, but also that he can see with his inner eye where Ahmeed has gone. In this process of becoming aware of space-time from within, so to speak, his power animal introduces himself to him by a distant, soft sound:

It intruded on my being, almost as if introducing itself: I am Tiger. It
was that sound between purring and growling that tigers make when
they are not sleeping and not hunting.

(Wolff 2001: p. 160)

What is interesting about the introduction of tiger is that he apparently
announces himself, as (Wolff later realizes) all animals that have sound do,
signifying: *I am here*.[3] After he returns in the village, after dark, he further
realizes that tiger has somehow accompanied him back home, from a distance. This falls in place when later Ahmeed asks him 'who brought you
home?,' to which Robert answers: 'tiger.'

The entirety of the initiation experience is described by Wolff as learning
what *learning* is (Wolff 2001: p. 169). Learning is here opposed to the
transfer of theories, methods and data, signifying instead the opening up to
one's surroundings to experience in a direct manner what they *have to tell you*.
This way of knowing, Wolff later finds, also makes him find certain medicinal plants when needed on a field trip. In his own evaluation he has stepped
outside the boundaries of the modern Western way of knowing, and has
found how the 'original' one of non-modern peoples can be re-activated.
Thus he finds proof of what an elderly lady, herself a shaman on the island of
Tonga, told him many years before, when he complained that so much of the
old knowledge and wisdom of humanity is dying out under the influence of
modernization. The woman comforts him, by answering:

'When we most need it, someone will remember that ancient knowledge
... traditions may be lost, but the information is in here and in here,'
she said, pointing to her head, then to her heart, 'and when we need it
most, it will be inside us, for us to find.'

(Wolff 2001: p. 6)

Now we may ask what kind of tiger it is, that presented itself to Wolff, or –
putting it differently – *how* the tiger was present. The question here is not
how a specific kind of communication can be explained within the frame of
standard modern science – the default question in modern time and culture
– but how and under what *form* tiger appeared – a question about the phenomenology of experiencing a power animal. What does the appearance or
self-presentation of the tiger tell us about the ontology at work, the ontology *taking place* in the jungle? It is significant that Wolff avoids choosing
between indicating a concrete singular tiger and the general species tiger, by
skipping the use of an article. *Tiger* presents himself, not *a* tiger, or *a specific*
tiger. In this spirited encounter with the jungle, where everything is abun-
dant with life and expression, the animal takes on a symbolic form so to
speak – being teacher and guardian to Robert Wolff. Hyper-sensory as the
experiences in the forest are, things don't seem to fall in the dichotomy of

empirical versus theoretical knowledge – of appearing as a singular or a universal. The real tiger is at the same time a symbolical tiger. A sign, signifying a way.

As much as Wolff crossed the imagined borders between the world of 'civilization' and that of the 'indigenous' Sng'oi, he still kept the oppositional thinking in terms of two worlds in place. He describes his shamanic experiences as leaving the modern world. When he takes up an assignment outside of Malaysia and his contact with the Sng'oi gets lost, he finds that his ability to enter the spirit realm diminishes again, except for rare situations when he really needs it. After his retirement he tried to return to an indigenous lifestyle on his own, living from birds' eggs, edible plants, roots he found on his piece of land in Hawaii – thus living a life which he had rediscovered and which he described as more 'human.' Before retiring, work, having a job, being part of the knowledge factory of the power system of modernity, impeded him to keep to his new-found indigenous life style:

> For many years I had to work so hard to do the things I was supposed to do that I became deaf and blind to what is important inside me. My luck was to find people who were human in an ancient way. My luck was to recognize and reclaim a humanity rooted in the earth.
>
> (Wolff 2001: p. 197)

Be swallowed

Malidoma Somé, the Burkinabe 'guru' who built a life trying to introduce Westerners to ancient African ways of relating to the world, like Wolff was led by life's events to consciously pass over from one way of being in the world to another – from a Western-styled worldview to a shamanic one. As a child (before decolonization) he was taken from his village by a priest who gathered young boys with potential from different peoples, to bring them to a Jesuit training institute, where they were made to forgot their language and their life with their families, to be remolded '... to create a "native" missionary force to convert a people who had wearied of their message along with their colonial oppression' (Somé 1994: p. 2). Certain memories of his early childhood however stayed with Somé, most importantly of his grandfather who had great spiritual powers and who had shown his grandson, as young as he was, certain things about his road in life. When a young man, finally fed up with the violence the teachers inflicted upon their pupils, Malidoma hits a teacher and flees.

Only remembering the name of his village, but not knowing how to get there, he starts walking, and, like in Wolff's case, the tiring journey without anything to eat brings him in an altered state of mind. When he is exhausted after a day's walk and sits against a tree to rest, a giant bird flies right up to him. He wants to hide but cannot move.

> The bird grasped me by the shoulders and ... took off with me hanging underneath. It was so dark I could not tell where I was being taken to, but I knew we were covering a great distance. Then, as if by enchantment, all of this vanished, and I found myself sitting on a soft, hairy body that somehow felt very familiar.
>
> (Somé 1994: pp. 143–4)

Upon waking from his dream, he finds feathers, hair and scratches on the tree. Wondering whether his experience is a hallucination or something real he has the impression that '...a powerful, nurturing presence lingered nearby. I convinced myself that something good was following me, looking out for me – that an ungraspable force was expressing itself around me' (Somé 1994: p. 144). From that point on, Malidoma gets used to the life on the road, finding fruits and water that sustains him, until he reaches home a few weeks later.

It is at that point that the most compelling part of the story starts – the narrative of a personal transformation, that helps the author to bridge the colonial world of the Jesuit school with the postcolonial and traditional one of his village. That his transformation is harder, and requires more, than that of someone just bridging the village and 'town' where modern jobs can be held, is clear from his depiction of the postcolonial reality that had come into being while he had been isolated in the seminary school:

> A new culture was born, the culture of a working man who would live abroad because of his work but who could return to his village if he wanted. I, on the other hand, had acquired something different and infinitely more dangerous: literacy. As an educated man I had returned, not as a villager who had worked for the white man, but *as a white man*.
>
> (Somé 1994: p. 167, italics A. R.)

This sentence contains the hard rock in Malidoma's life – as much as he has tried to recapture Dagara culture, through his forced colonialist schooling whiteness had become a part of him, and, it may be argued, has never left him since. In such a situation bridging cultures, and their ontologies, seems impossible. The impossibility however reflects the order of rational logic and its law of the excluded middle: one participates in whiteness, or one doesn't. In the Dagara logic, as Somé describes it, his condition is not so much a thing to be erased in order to become a member of the village again, it is to be conjured, so that the inherent danger of whiteness may be contained:

> People understood my kind of literacy as the business of whites and non-tribal people. Even worse, they understood literacy as an eviction of a soul from its body – the taking over of the body by another spirit ... the ability to read, however magical it appeared, was dangerous ... To read

was to participate in an alien form of magic that was destructive to the tribe. I was useful, but my very usefulness was my undoing.

(Somé 1994: pp. 167–8)

The presence of the spirit of the whites, and its inherent brutality and violence, made it a question of doubt whether Malidoma could still be taken up into the tribe through the rituals of initiation. In the end it is decided he will be accepted to undergo them, even though it is highly uncertain whether he will survive the magic happening in the process – even more than in a normal case.

The descriptions of the initiation process form the heart of the book, and are hallucinating to read. In them Somé renders his experiences of his surroundings in altered states of mind as if they are experiences of a world as real and touchable as the one of everyday sense experience. To give just one example of pages filled with descriptions of ecstatic experiences, I will cite from his report of the first night at the initiation camp.

All around me and underneath me I could feel life pulsating, down to the smallest piece of dirt on the ground. The way this life expressed itself was otherworldly: sounds were blue and green, colors were loud. I saw incandescent visions and apparitions, breathing color amid persistent immobility.... Each person was like the sum total of all the emanations taking place. The people, however, were not in charge of the operation of the universe around them – they were dependent on it and they were useful to it as well.

(Somé 1994: p. 201)

The 'hyperreal' style seems to be his conscious choice as a writer, and is consistently adopted in all his books. It may be one of the reasons that his autobiography became so well-read, attractive as it is to Westerners seeking for more intense ways to experience reality, be it through mind-altering substances or otherwise. In fact Somé seems to have adopted a strategy for which the missionary Jesuits who trained him have been famous – to first get to know as much as you can about the culture you enter: to learn the language, to get to think like the people you want to convert, to get into their skin and their mind. From that point you can start to describe the world in their idiom, so they can take the worldview you are offering them as reality, instead of as something to accept through a decision of faith. In the process, however, the missionary message will have changed as well. It will have acquired another cultural flavor and taken up elements of the ontological idiom of those to be converted. This missionary method could be described as a method of ontological *fusion*.[4] Once you have taken in this fused worldview, you cannot dispose of it again for fear of losing your identity which now has been swallowed by it. What Jesuit missionaries did to South

American Indians, to Africans and Asians, Somé does as a Dagara to his Western audience – convert them to a shamanistic ontology which is not purely traditional Dagara, but which is fused with the modern (American-puritan?) preference for direct experience to convince the lonely doubting mind, over things valued more in traditional cultures, such as authority of the elders, family relations, etc.

After his initiation into Dagara manhood, years later than usual for boys in his village, Somé is sent out by the elders to go and study and live 'out there,' in order to fulfill the name that was given to him before he was born: be friends with the stranger/enemy. After studying in Burkina Faso and in France, he became a lecturer in the US, a position he eventually gave up to become a shamanistic coach and writer. In this function he organizes trainings for Westerners, mostly Americans, to reconnect with the spirit world. In his book *Ritual: Power, Healing, and Community* (Somé 1993), he presents his Dagara critique of modern life, as well as ways to remedy its wrongs. Modern life is described by him as a life of acceleration and losing one's soul to 'the machine,' the corporation and its anonymous power. The loss is on everyone, the wealthy as well as the poor, as community is destroyed:

> ...behind the mighty-looking corporations are a group of wealthy people whose personal lives are lived in marginality ... These people start to become invisible because they are mere instruments of the power being displayed, the power being made visible. They take a back seat to the corporation's need to be powerful.
>
> (Somé 1993: p. 41)

And:

> It is the action of those in power that produces the poor, the menial worker, the man and woman in debt and the homeless.... The menial worker, the man and woman in debt, the poor and the homeless exist, as if they must, to highlight the person in power.
>
> (Somé 1993: p. 41)

The cure to this loss of soul is simple, according to Somé: if one refocuses on *ritual*, the soul, the true purpose of life and the sense of community will come back and be restored, for

> ...whenever ritual happens in a place commanded by or dominated by a machine, ritual becomes a statement against the very rhythm that feeds the needs of that machine. It makes no difference whether it is a political machine or otherwise.
>
> (Somé 1993: p. 19)

As a shamanistic coach in the USA, Somé started experimenting by organizing rituals in group sessions that were adapted to the psychological and social existential condition of Americans. He has received much criticism from different sides for his idiosyncratic approach – for not representing 'true' Dagara culture, and for being the instrument of just another self-serving sectarian movement. In his books however, he shows enough self-doubt to accuse him of false intentions. He concludes his book on ritual by asking himself whether his aim '...to impart to the modern world that which is rooted in the ancestral world...' will be attainable (Somé 1993: p. 102). Referring to the collective unconscious of Jung he expresses his hope that deep spiritual knowledge can be shared across cultures:

> And so I offer the prayer to our common ancestors on behalf of those seeking to recover themselves from the rubble of modernity as they seek to work their way toward being elders of the new post-modern tribal order.
>
> (Somé 1993: p. 103)

At the end of his autobiography he expresses even more caution, upon describing his mood when in college in Burkina Faso:

> My enduring passion for magic, rituals, and ceremonies reassured me that the traditional world had swallowed me and that I was resisting the white world – or maybe I had grown to be a man trapped between the white and the traditional worlds.
>
> (Somé 1994: p. 311)

In the end it is his trust in living out his personal destiny that sustains him, following the words of his friend and guide Guisso 'Go and allow yourself to be swallowed' (Somé 1994: p. 311). Only by letting himself be swallowed by the *Western world*, the seed of Dagara heritage he bore in himself would then be able to sprout there.

Indigenous versus modern – conceptual issues

In both stories we see their authors struggle to bridge what they at first experience as a chasm between a rationally ordered 'modern' perception of things, and a traditional, 'ancient' way to be with all kinds of energies, be they ancestors, plant or animal spirits, or whatever kind of spectral beings that invite themselves to their dreams. I chose these authors to introduce what is at stake in preparing for a dialogue with shamanistic ontologies, because in both life's stories we see how a new way to experience went together with the realization of a new commitment to healing relations – may they be imperfectly perceived, or restricted by the constraints their lives put on them. Also both authors clearly indicate an awareness of what is wrong in modern ways to relate to nature, and a commitment to indigenous ways of life to repair what is lost to the moderns, and through them to the world. Let me elaborate on these points somewhat more, meanwhile clarifying some of the terminological and methodological issues that are at play in this book.

Several things need to be mentioned here: first that a book like the present one, which addresses a big question (negotiating the environment) necessarily zooms out. It can only speak about the things at hand in tentative ways, and can not make a claim to the most precise definitions of the main concepts. To just mention one thing: when I speak of 'spirits,' the reader may think this is a simple category with a clear reference, for instance: immaterial energies that can only be perceived in special conditions, like in induced trance. It is not that simple, however; one cannot just give a definition and move on. There are innumerable studies of spirit beliefs, spirit experiences, the function of spirits in societies, etc., which are written from different methodological viewpoints, such as sociology, psychology, anthropology, history of ideas, religious studies. Besides, there are first-hand reports of shamans and people from shamanistic cultures, which are the sources that should interest us especially. All of these sources may use the word 'spirits,' and refer to different realities, experiences and methodological and phenomenological frameworks. I hope and aim to find a way between all these discourses and methodologies, aiming at an interdisciplinary philosophical reflection, which necessarily has to bypass the hopes to form a unified theoretical and conceptual language of spirit realities. In order to meaningfully address big questions, one has to put methodological purity second. Meanwhile, however, one must try not to sacrifice clarity and preciseness.[5]

When we focus on spirit ontologies, meaning more or less stable realities that count spirits in and that are the basis of life in traditionally living indigenous communities, we will also see many different *indigenous* descriptions of such realities. Those descriptions differ among themselves, and hardly harmonize with modern Western conceptions of spirits. To clarify this, let us look into one of the rare books in which a French anthropologist, Bruce Albert, and an indigenous shaman, the Brazilian Yanomami Davi Kopenawa, have made an effort to bring a particular indigenous (Yanomami) spirit ontology across. Kopenawa addresses his anthropologist friend Bruce Albert in the very beginning like this:

> Your professors had not taught you to dream like we do. Yet you came to me and you became my friend. You put yourself by my side and later you wanted to know the words of the *xapiri*, whom you call spirits in your language.
>
> (Kopenawa and Albert 2013 [2010]: p. 11)

The important part in this quote is 'whom you call spirits.' This calling does not at all guarantee that the same reality is addressed. It only says that a word is chosen, 'spirits,' which comes closest to indicate what is at stake in the Yanomami reality. Albert, who helps the modern reader with an impressive amount of footnotes, to not interrupt the voice of Kopenawa, adds in footnote 3, p. 490:

> Any existing being has an 'image' (*utupë*) from the original times, an image which shamans can 'call,' 'bring down,' and 'make dance' as an 'auxiliary spirit' (*xapiri a*). These image-beings ('spirits') are described as miniscule humanoids wearing extremely bright, colorful feather ornaments and body paint.

This description may reduce the default modern reader's confidence that he can simply touch on the reality of 'xapiri' by using the word 'spirits.' The aspects of dancing, of their bodily appearance as self-adorned humanoids and possibly other aspects, already got lost in the modern philosophical notion of a spirit as a disembodied entity.

Further complexity is added when we read that certain groups of Yanomami also use the word *xapiri* for shamans. Doing shamanic work is probably understood as acting as spirits do. Albert explains that this indicates how the shaman in his trance identifies with the helping spirits he has summoned up. Now were we to investigate descriptions of how spirits are real in indigenous communities all over the world, from Siberia to Australia, and from America to the last remaining ones in Europe, not to mention those still surviving on the vast African continent, we would see a plethora of rich and differing descriptions, and we could loose ourselves in pointing out all these differences with great precision. All the same, in our present globalizing times, most indigenous peoples are represented by (unofficial) ambassadors and activists, like Davi Kopenawa, who connect with each other as well, recognizing the commonalities of their struggles against institutions and businesses that destroy their habitat and ways of life, but also the commonalities in their ways to relate to nature, to take certain things to be real that modernity does not and to consequently relate to nature in ways that are contrary to the goal of modernity to 'master and posses' her.

In this book I will not try to reconstruct a single generalized 'shamanistic' or 'indigenous' worldview or way of life, to oppose it romantically to the evils of modernity. Such an opposition runs the risk of remaining bound to the same framework it seems to counter – to present just another 'consumerist' grab at indigenous ways of living as a medicine for the ills of modern society, while not critically engaging with the political-economic power systems that endangers such life in the first place. Such a critical engagement can only take place by leaving the task of expressing and describing the realities of indigenous cultures on principle to their shamans, thinkers and speakers, restraining oneself to not produce knowledge of the realities they live in and perceive. The aim of this work is a different one – to focus on where these realities and the realities of modernity meet and clash, often with deadly and wounding results. To focus on a philosophical articulation of how these clashes and encounters have come to be, in their historical and existential situatedness, through the spreading of *political epistemologies* that favor modern ontologies of nature as dead and dispirited, and which in turn enable the actual destruction of actual people, animals and other natural beings, through state and corporate activity, ideological aggression (religious conversion) and legal and non-legal, state and non-state interventions in communities on the ground. Philosophy does not seek primarily to describe and explain as empirical sciences do, but to articulate the imaginary constructs through which we, human beings, relate to our world. Constructs that are not imaginary in the sense that they are 'made up' and unconnected to what is given to us, but in the sense that they consist in interconnected constructions of 'images,' which are created in order to collectively deal with the world we humans live in. Just like the law is a collection of 'images,' descriptions of how we perceive just relations should be: descriptions that guide our actions and that work on our actual relations. The law has reality, while being 'imaginary' at the same time. In this sense our culturally varying ontologies describe how we *perceive*, in varying conditions, that *our relations to nature in the most fundamental sense should be*. They guide our actions and work on our actual relations in the most general sense. While we articulate such constructs, we can evaluate how they do it and whether the perceptions they express are true to what we, humans, can and should be and do.

Such a critical work is, naturally, a historical work. It has to reckon with how the conflicting ways to relate to nature have come into being and have actualized in historical processes, which are complex, not uniform, ridden with internal tensions and with opposing tendencies playing out in them. This means that indigenous ontologies are not, like romanticizing views have often depicted them, ancient, original, natural, unchanging, the same for thousands of years or the same as those of our early human and non-human ancestors. They are actual historical, changing, locally varying answers to the conditions which the peoples that acknowledge them perceive their relations to nature should be. The same holds for modern ontologies – they neither represent one, single, essentially unchanging ideology or way of life, but rather the varying answers by different peoples and groups to complex historical processes. The articulations of how modern relations to nature should be are not one and the same in the seventeenth and in the twenty-first century, nor were they ever the same in the cities and in the countryside, or in Albania or California or Prussia, to mention just a few different places in what we call the 'modern' world. Ontologies are nothing more or less than *relatively stable relations to the world we live in*, which are continuously negotiated in thinking and in practical activities of members of relatively stable groups we call peoples, communities or cultures. We can only speak of such general categories as indigenous versus modern by zooming out, and because I aim at the critical work of philosophy, I will leave the relevant differences between cultures, groups and situations covered by the categories of modern and indigenous for others to describe and define in more detail.

Now when we turn to the word 'indigenous,' this creates its own complications. It is a word that was introduced by the very colonial world that forms the historical frame of the issue at hand (cf. Müller 2013: pp. 12–13) – the world that ventured out from Europe to 'discover,' conquer and take in possession the worlds beyond European lands. The colonial project preferably described the lands it would take under control as empty, wild nature. If it found people living there, it tried to describe them as not fully human, as uncivilized, almost identical with that same passive nature, while it reserved history, agency and civilization for itself; thus allowing the ideological defense of the enslavement as workers of colonized people as well as the destruction of their civilizations. Although the word 'indigenous' may sound innocent enough, its venom is in the fact that it *others* people as non-modern, non-global, it localizes them, connecting them to the places where they are 'found' by the European explorers/colonizers/travelers – over against the Europeans who see themselves as modern, as having a natural right to go around the earth and be citizens of the earth itself and conquer it. This denies that all human beings have a migratory history, that being human itself means adapting to natural conditions in ways that may involve moving and being principally at home anywhere in the world. Despite all of its problems, however, I still choose to use the word indigenous. The main reason for doing so is that the shamans and activists of the peoples concerned, who nowadays travel, communicate their values to the world and do legal and political representation on a global scale, use it themselves. They use it to indicate how they perceive their people's desire to lead their lives differently than the moderns do – not focused on money, wealth or consumption, but focused on how the spirits tell them to maintain an environmental balance with specific local surroundings, out of respect and love for the actual living beings that make up those surroundings, on which human life depends, and which human life can help to sustain.

Similar observations can be made about the concept 'shamanistic.' In modern Western culture it is common to think that human and non-human living beings inhabit ontologically distinct spheres – those of culture versus nature, of reason versus instinct, of values versus facts or of spirit versus matter. Such an understanding of the world, it is clear, is not universally shared. In what I will call shamanistic cultures, humans and other living beings are viewed as taking part in shared or overlapping ontological spheres, in which all are understood to interact, depend on and communicate with each other. The continuous communications and interactions in a shamanistic reality will not all be reflectively noticed. They may be back-grounded in everyday practices such as resting, eating, planting or hunting. In certain emotional situations such as fear, attraction or joy, they may be centered or foregrounded. Finally, certain communications and interactions, like those with spirit animals, the spirits of mountains or plants, are most notably experienced in special situations, especially in trance states, that are considered to be the specialty of shamans. In this work I will follow I. M. Lewis, who in his 1971 study *Ecstatic Religion, A study of Shamanism and Spirit Possession* (2003 [1989]), rejects the anthropologically purist choice which preserves the word 'shaman' only for those spiritual mediators that were called that way in their own language – the Tungus in Siberia. Like Lewis, I will use the words 'shaman,' 'shamanic' and 'shamanistic' as more flexible and extended terms: 'I prefer to see shamanism as a general, cross-cultural phenomenon based on the shaman's mastery of spirits and the practice of his art with the aid of spirits'[6] (Lewis 2003 [1971, 1989]: p. xix).

Elsewhere I gave the following argument for a cross-cultural use of the term 'shaman':

> Concepts are extended in their use, re-used, even recycled and put to new use all the time. Just as in the second half of the twentieth century, the Indian term 'guru' came to signify any kind of spiritual teacher, so the term 'shaman' nowadays extends to all kinds of spiritual mediators across the globe – those who work in old and mainly local traditions as well as those who experiment in nonlocal ways with practices from those traditions. This new use of the term 'shaman' is so much already a reality that we should just go along with it.
>
> (Roothaan 2015: pp. 141–2)

When people from modern, more secular, environments think of shaman-ism, they tend to focus on the alternate consciousness[7] that shamans enter when going into a trance state. The extraordinary actions of a shaman, pref-erably dressed in traditional clothing adorned with animal feathers and amulets, when (s)he goes into trance to find healing, prophetic or life-guiding knowledge, incite awe and curiosity among the moderns, who see something happening which their science-based worldview takes to be impossible or nonsensical. To those living in shamanistic or indigenous cul-tures, such special, 'paranormal' experience is not something to revel about, but rather taken as a life-saving practice, whenever guidance or healing is required. A practice that can be trusted because it takes all the ontological commitments of daily life (commitments to the interconnectedness of all life in and around us) serious. It is part and parcel of a wider way to be in the world, in which all kinds of things – stones, sunny patches on leaves, frowns, smiles and frog's eyes, may have a significance for life, and may communicate that to us. In modernist orthodoxy the awareness of the interconnectedness of everything in our world, *and* cross-species, extra-human communication, have been filtered out, while valid experience is reduced to the framework of science (causality, chronology, space-time). In the following chapters, however, I will not treat modern culture, and modern thought to be of a single stock. The idea of a single, hegemonic, way of being in the world that is to be called 'modern' is to be rejected. This rejection is supported by a discussion of marginalized parts in modern Western history of ideas in Chapters 3 and 4. There it will be shown that even in a modern context the outlawing of the spirits has always been contested, as well as philosophically problematic. There is no way one can theoretically confine spirit ontologies by identifying them as ancient and the clear and distinct opposite of modern, secularist ontologies. Before going there I will however first, in the next chapter, investigate recent attempts in cultural anthropology to go philosophical and to try to end the othering of indigenous knowledge by means of a decolonization of anthropological research.

Notes

1 Despite Goodman's recognizable experience, it should be noted that even in highly technological societies nature is not completely disenchanted however, as we notice for instance when tunnel drillers place a statuette of a saint before the decisive part of the work is started.

2 Whiteness is, just like blackness, a word of power – not a concept that refers to some empirical reality, but instead defining the empirical. It should be clear that there are indigenous peoples with light skin color, such as the Saami in northern Europe, and there are people with dark skin color who take part in the power systems of modernity. What whiteness and blackness as concepts convey is the dominance of racist imperialistic cultures and corresponding ontologies and knowledge systems that arose in Europe and have globalized and transformed all cultures of the world.

3 In the chapter on the elephants we find the same experience, described by Lawrence Anthony – who also learns to sense the herd coming near before hearing, smelling or seeing it.

4 The common name for the Jesuit missionary methods is cultural adaptation or accommodation. A historical overview of the development of these methods, with a focus on Asia, can be found in Pina 2001.

5 This approach is defended convincingly in Gordon 2006.

6 The history of how anthropologists and sociologists of religion have looked at shamans and shamanism will not be discussed here in depth. For the concerns of this book I start out with much of what Lewis argued, doing away with ideas of older authors such as Frazer and Eliade, whose descriptions of shamanic trance as 'celestial voyaging' rather reflect a Christian theological outlook and do not fit self-descriptions of shamans of their experiences. Lewis introduces a sociological approach, based upon empirical work. My purpose moreover is not to describe and understand shamanic trance states, as is the aim of Lewis and later researchers of shamanism, nor to understand its sociological functions in indigenous societies, but rather to explore and express philosophically the understanding of the world that is opened up in trance states, and which vice versa makes these trance states meaningful.

7 This term was introduced by Pieter Craffert, to avoid the use of 'altered' state of consciousness, because it is impossible the give a cross-cultural definition of what 'altered' is: 'What is "ordinary" consciousness is not the same for all human beings, and on a cultural level, a distinction can be made between baseline (or normal) and alternate states of consciousness, and they will differ from culture to culture' (Craffert 2008: p. 23n22).

Chapter 2

Ending the othering of indigenous knowledge in philosophy and the ontological turn in cultural anthropology

Back to the phenomena

There is no single modern philosophy. Philosophy is always plural, and tentative, wherever it is found. It is dubious to speak of modern culture, or modern philosophy and science in the singular, just like it is tricky to speak of shamanistic cultures or shamanistic philosophy as being of a single type. In the previous chapter I explained that this is one of the consequences of zooming out, a move that is needed if one wants to address interconnected issues as this book aims to do. Generalizing concepts are also functional because they influence people's thought and action, and thus have pragmatic reality. Still, the mentioned generalities hold together widely diverse systems of thought, life practices and ways of knowing things about the world. One of the things I hope to show in this book is that modernity, in its generalized sense, should be understood to have involved simultaneously the discovery, by European peoples, in their own context, of new intriguing values such as autonomy, social justice and emancipation, as well as the strong drive to colonize the earth and non-European peoples. This colonization meant that all these 'others' had to be silenced as best as possible, which created a strong tension and ambiguity with respect to the liberating aspects of modernity. The ideological narratives and oppressing structures meant to silence the others have also continuously been fought by those who were oppressed and subdued, but we find a continuous stream of criticism of the silencing ideology of modernity among (non-mainstream) thinkers in Europe as well.

In this chapter I will discuss several strands of academic work (some of it combined with activism) that have challenged modernism from within. The most recent ones are the movement that has adopted 'animism' as a self given label, and the so-called ontological turn. The first-mentioned is to be found among eco-feminists, environmental ethicists and researchers of religion (Harvey, Plumwood, among others), and stresses the need for a change in attitude and morality among the moderns, in solidarity with and also listening to indigenous peoples criticizing modernity. The other is to be found among anthropologists (Kohn, Viveiros de Castro, among others who seek to decolonize their own discipline by taking the ontologies of indigenous peoples seriously, albeit using Western philosophical systems to articulate them. Both these strands of work rely, philosophically, on ways of thinking that make it possible to reverse dualisms that have become characteristic of modernism. Plumwood called such dualisms 'hyperseparation':

> ...an emphatic form of separation that involves much more than just recognizing difference. Hyper-separation means defining the dominant identity emphatically against, or in opposition to, the subordinated identity, by exclusion of their real or supposed qualities. The function of hyper-separation is to mark out the Other for separate and inferior treatment.... Colonizers exaggerate differences ... The colonized are described as 'stone age,' 'primitive' or as 'beasts of the forest,' and this is contrasted with the qualities of civilization and reason that are attributed to the colonizer.
>
> (Plumwood 2003: p. 54)

These words explain what 'othering' is about – to be marked as 'other,' with the effect that one can be treated differently and worse than the group that has the upper hand. The concept of othering is used as a critical tool in feminist work, and also in decolonial studies. The one who does the othering, the 'self,' who is the measure of identity, of normalcy, intelligence and of the right values, makes himself invisible. That one, 'man,' is not gendered, is not sexualized, has no color, if we believe him, no age or other embodied features – he is just the seat of pure reason and goodness.[1]

Othering can also take place as 'backgrounding,' another concept often used by Plumwood. Those who are pushed in the background should automatically and willingly take it upon themselves to care for and serve 'man,' to make him look more on top, more interesting, more unique, when he is exploring and inventing things. The eco-critique of modern thought and culture that Plumwood initiated, roots in an Australian setting – from her Aboriginal teachers she learned that locality is decisive for life as well as for understanding. In Aboriginal ontologies *the land* is seen as a primary agent, birthing and steering all that moves and lives on her and off her. To follow Plumwood and the other eco-animists one should try to convert the 'normal' Western interpretation of the Aboriginal speak of agency of the land as a poetic addition to factual reality. It can be seen the other way around: seeing the land as a passive, non-living entity, as a background for biological activity, is the effect of the colonizers gaze, that conjured the land's agency away:

> The colonizing framework's exclusion of the non-human from subject status and from intentionality marginalizes the non-human as narrative

subject and agent, and pushes the more-than-human sphere into a background role as a mere context for human thought and life.

(Plumwood 2003: p. 66)

Graham Harvey has tried to make that conversion as well. In his book *Animism*, we read in his chapter on Aboriginal Law and Land:

The way things are arises from the lands. The places where things happen are not mere scenery, backdrops or stages for the great drama of life; they define, birth, contextualise and participate. Lands erupt into life and fully engage with the emergent and proliferating diversity.

(Harvey 2017 [2005]: p. 65)

A description follows of what he learned from a trip to Alice Springs with his Warlpiri guide Bob Randall.[2] When they get out of their car at a rock identified as the entrails of a dog that lost a fight with another,

Bob said that life erupts here. It was here, most clearly, that I realized that Dreaming is not well represented by gentle Just-so stories about the emergence of places, as if someone placidly painted the landscape onto a blank canvas. The lands were, are and will continue being formed by the eruption of that which was not here.

(Harvey 2017 [2005]: p. 77)

This attempt, by a white author, trained in modern Western scholarly work, to find language to express the Aboriginal world *without* othering it once more, is interesting. Harvey, who aims to leave behind the dualisms of Cartesian philosophy and Western Christian theology that formed our modern discourse, of course cannot do so entirely, when he summarizes ' "Creation" is happening now as the Law is abided by in the lands' (Harvey 2017 [2005]: p. 68). Here we stumble at one of the major issues that arise in striving to speak with indigenous peoples, instead of about or even for them – the issue of an inescapable untranslatability of discourses as they articulate different ways to be in the world.

To dig into the issue of translation, and other problems connected with the impossibility to fully enter another's discourse, or to bracket one's own, one needs the reflective tools of intercultural philosophy. From the beginnings of this new branch of philosophy,[3] hermeneutics of dialogue and deconstruction have been adopted as its major approaches. These approaches focus on translatability, understanding and (mis)communication. An intercultural hermeneutics – presupposing two or more parties that have to be respected in their own right – has to leave the idea of a 'fusion of horizons' as articulated by Gadamer. Focusing on dialogue all the same, it cannot make use of a hermeneutics of identity either, and always already will be in the process of deconstruction. In the words of Mall:

> Intercultural philosophy ... does not unnecessarily give privileged treatment to any philosophy, culture, or religion ... For this reason, intercultural philosophy rejects the idea of a hermeneutics of identity that is intolerant of difference.... It approves of overlapping centers, searches for them, finds them, and cultivates them.... The modern-postmodern debate thus loses its sting when put in this context of overlapping and abiding structures.
>
> (Mall 2000: p. 6)

In this book we will need to move beyond Mall, who seems to think that overlapping centers can somehow create dialogue without meeting with too much conflict and struggle. The reason is that, apart from the necessity to also investigate the power systems that have suppressed the voices of the others, we need to also become aware that understanding (like Kant's pigeon) needs resistance to get off the ground, if we want to begin to understand cultures different from our own. Deconstruction points to obstacles, to silent inconsistencies, hidden presuppositions of any textual pattern, inconsistencies and obstacles that will create misunderstandings. These misunderstandings however will double as open spaces, as invitations, to give room to what is told by the other.

A hermeneutics of identity is self-contradictory, as hermeneutics already points to the horizon of any actual being in the world, and of any potential understanding. A horizon, as limiting boundary, already embodies the resistance needed for the movement that understanding consists of. 'Horizon' not only indicates the actual horizon of the place where I am, but also the historical, cultural, political and economic conditions that frame my mind. When different ways to speak of reality meet, when ontologies bump into each other, the pre-given structures of reflection of the one will have to be bent and stretched to give way to some of the insights of the other. So 'creation' (in the above quotation of Harvey) can indeed be used to clarify to a Western reader what the entrails of a dog in Alice Springs are about, as the choice of this word is not more or less than a movement to try to understand the Aboriginal discourse. The original inhabitants of Australia (and of any colonized land) will of course have had to do similar acts of interpretation to try to understand their oppressors, for the dual purpose to defend their lives and way of life, and to understand what this new 'culture' was all about. Since colonized peoples thus already practiced intercultural philosophy as a survival strategy, long before it became an academic discipline, we have the more reason to understand intercultural *dialogue* as only the last step after deconstruction, criticism and negotiation of what is oppressing and conflicting in the encounters at hand.

Hermeneutics, deconstruction, showing attempts to resist the 'dualisms' of modernism, now developed their reflective paths from phenomenology, which can be characterized as the most influential movement in Western philosophy to turn early modern dualisms around. Husserl's intentionality, for instance, aims to 'think together' subject and object (the subject that is already with the object), like Heidegger's existentials aim to highlight that any possible thought is already an answer to a condition of care for being-there (dasein), and James's pragmatism aims to characterize human epistemic convictions within the larger process of adapting to and changing the world to an environment that makes human beings thrive.[4] Phenomenology resists the modernist divide between sense impression and reason, body and mind, feeling and distanced observation, and it does all of this from the proclaimed need to center phenomena, or appearances. Appearances are here not opposed to objective states of affairs, or supposed to come to us in pure sense impression, but are seen as the original opening up of the world to 'us' (humans who perceive). Pure and objective sense impressions (like: I see that this ball is red) are considered to be reached only after abstracting from the original 'mixed' (or 'rich') way in which things appear to us normally. In the words of Merleau-Ponty:

> Instead of providing a simple means of delimiting sensations, if we consider it in the experience itself which evinces it, the quality is as rich and mysterious as the object, or indeed the whole spectacle, perceived.
>
> (Merleau-Ponty 2002 [1945]: p. 5)

> The traditional notion of sensation was … a late product of thought directed towards objects, the last element in the representation of the world, the furthest removed from its original source, and therefore the most unclear.
>
> (Merleau-Ponty 2002 [1945]: p. 12)

For the purpose of the present investigation I propose to understand phenomenology as the urge to re-root human understanding after modernism. Phenomenology doesn't do away with modernism, it saves it, and its most important tenets.[5] It tries to understand modernism, however, as just one more cultural approach invented or discovered by humans in their dealings with life. It situates it historically, and recognizes that it demands a commitment to its values in order to endure. As much as phenomenology thus aims to overcome dualisms, it has grown out of the same epistemological network that brought forth the very dualisms it criticizes. The epistemological network that sooner or later always starts to mention Plato or Socrates, that slides over 'innocent' racisms and that does not realize that it is always (still) othering women, non-whites, disabled people, children, and down the pyramid of worthlessness, animals, vegetative and mineral life. A true intercultural philosophy cannot be realized without the development of real engagement with the consequences of the dark sides of modernism, its power grasps, its colonizing and hegemonizing actions. Such an engagement cannot halt at a 'decolonization of minds' (as important as that is, in itself), but will be willing to be judged by those it othered and to accept the verdict and demands for retribution they will bring to any type of white philosophy.

A special place in phenomenology in the wider sense is taken by William James, whose work I will discuss in more detail in Chapter 4. James was also deeply rooted in modernism, as is clear from all the time he spent to build the foundations of psychology as a science. For long years he worked on one of his major works, the two-volume *Principles of Psychology*, which became a standard text book in the study of psychology. All the same, with intervals but during his entire adult life, he was also involved in the search for foundations of psychic research. In her 2006 book *Ghost Hunters* Deborah Blum has detailed James's engagement with investigations into psychic phenomena, as well as with the founding of the American as well as the British Society for Psychical Research. These societies believed in the necessity and possibility to carry out scientific research into phenomena of telepathy, telekinesis and mediumship. The promise of newly developed technologies such as telegraphy and telephony made interested scientists expect that instruments could be developed that would measure the psychic energies involved in such phenomena. The work of the societies also included debunking certain psychic performances, among others of the then famous Madame Blavatsky. Debunking was however not their intrinsic aim, it was just part of the work to try to prove the workings of those psychic processes that were not part of public performances, of circus-style miraculous acts, but that seemed to show true contact with the spirit realm.

Next to this scientific research into psychic phenomena, James built on the work of the panpsychist Gustav Theodor Fechner (1801–1887) to create a philosophical understanding of spirit phenomena as inherent aspects of the experienced world. In fact he combined his work in such diverse fields as psychic research, 'normal' psychology, philosophy and the study of religion to this effect. In order to rethink our relations to nature in such a way as to include spiritual matters, James held a better understanding of the way in which we experience the world to be adamant. In a much more radical way than Merleau-Ponty, he propagated that we understand theoretical and empirical research (science) as a very specific, learned approach to the world, an exceptional way to go at things, quite distinct from the normal, everyday relations we have to our world.

> The original sin, according to Fechner, of both our popular and our scientific thinking, is our inveterate habit of regarding the spiritual not as the rule but as an exception in the midst of nature.
>
> (James 1996 [1909]: p. 150)

James aimed to introduce the little known Fechner into philosophy for his 'thick' description of the universe as being alive, spirited through and through. He appreciated Fechner's work above the American transcendentalists who were so well-known in his time and place. Philosophy, according to James, should try to express life, and not lag in rationalistic abstractions, be they of the idealist or of the realist type. He didn't agree though with Fechner's claim that each individual consciousness is like an organ of one all-encompassing consciousness. Against the strong monist tendencies among spiritual thinkers throughout history (from Proclus to Teilhard de Chardin) James held the very different view that spirit reality is *distributed and not unified*. In Chapter 4 I will argue that especially this view holds a promise for respecting (instead of othering) shamanistic ontologies. It makes it possible to take spirit ontologies of different kinds serious, and to include them in philosophical discourse without reducing their difference in any way. For now it is important to have a first grasp of how a more inclusive and non-dualistic philosophy grew out of the phenomenological critique of modernism, as it paved the way for the work of thinkers from a variety of disciplines to rethink our relations to nature, *and* to indigenous way of being in the world, in our time.

One among other animals

There are many persons that are not human, and we should relate to them with respect – that is the shortest summary of the non-dualistic position defended by religious studies professor Graham Harvey. This position is turning one of the most central ideas of modern philosophy on its head, the idea that 'man,'[6] the category used in classical modernisms to refer to human beings, has a unique place in the world. Before going into Harvey's provoking defense of a new animism, let me describe the relevant outlines of modernism's two-tier exclusion of animist ideas, referring to and adding to his discussion of such exclusion. A most prominent form of it can be found in the works of Enlightenment philosopher Immanuel Kant (1724–1804). According to Kant, man thanks his special place to his potential for reason, which he may perhaps share with creatures unknown to us (Kant presumes so much, allowing the potential being of intelligent aliens), but which distinguishes him here on earth from all other living beings. It also divides him *within* – alienating him from his animal side. With the special place mankind holds in the world comes a special dignity and right to protection, which other beings lack. Other life forms are in a sense 'dead' – even if they have biologically explainable urges, they can never be said to show *action* in the moral sense of the world, as action implies will and reason. Except for man, according to classical modern philosophy, no being except the rational being can be a moral subject, nor be considered to have real knowledge about the world.[7]

In *Animism*, Harvey cites that other famous Enlightenment philosopher, David Hume, who declared that any ascription of action to non-rational beings has to be seen as either poetic metaphor, or as ignorant stupidity:

> There is a universal tendency amongst mankind to conceive all beings like themselves, and to transfer to every object those qualities with which they are familiarly acquainted, and of which they are intimately conscious.... Hence the frequency and beauty of the prosopopoeia in poetry, where trees, mountains and streams are personified, and the inanimate parts of nature acquire sentiment and passion.... No wonder, then, that mankind, being placed in such an absolute ignorance of causes, and being at the same kind so anxious concerning their future fortunes, should immediately acknowledge a dependence on invisible powers possessed of sentiment and intelligence.
>
> (Cited from Harvey 2017 [2005]: p. 5)

The silent presupposition here is that human beings alone have consciousness and agency. But *among* human beings Hume also makes a difference: those who hold trees, mountains and streams to have personhood, and who recognize the reality of non-physical agents (spirits) – 'animists' – are called ignorant. The intelligence that provides human beings with the right kind of self-understanding is not evenly distributed, says Hume – for only the whites have shown the ability to develop the civilization based upon such discrimination between objective reality and poetry. In Eze's 1997 book on *Race and the Enlightenment*, in another context, we find Hume declaring

> ...the negroes and in general all other species of men ... to be naturally inferior to the whites. There never was a civilized nation of any other complexion than white.... No ingenious manufactures amongst them, no arts, no sciences.
>
> (Cited from Eze 1997: p. 33)

When we relate Hume's view of human races to his view of spirit belief, we must conclude that he denied that the ascription of consciousness and agency to non-human beings made any sense. To him any kind of animism is ignorant, and naturally present among racially inferior, non-civilized peoples.

The heritage of such thinking is still present in the work of early anthropological writers such as Frazer (1854–1941) and Tylor (1832–1917), who added to the racism of eighteenth century thought the idea that peoples who held animist views were in an 'earlier' stage of development. Evolution has become an important concept, since the publication of Darwin's *Origin of Species* (1859), also in the 'study of man,' anthropology. Where before Tylor it was considered only as a branch of philosophy, centering on the reflection of what it is to be human, now next to the 'white' sciences of sociology and psychology, which studied mankind in its Western variety, anthropology came to refer increasingly to the study of non-Western cultures. Now humanity came to be measured in relation to culture, and cultures were ordered on a scale of progress with regards to manufacture, trade, technology, science, religion and philosophy. Tylor, who is generally seen as the founder of cultural anthropology, '...was particularly concerned with "intellectual" development...' (Harvey 2017 [2005]: p. 6). He held animism to be a primitive human theory about the world, that projected a spirit or soul into everything. The 'mystery' that not all of humanity had evolved away from such primitivisms is explained by Tylor by means of the concept of 'survivals' – remnants of lower, earlier levels of civilization that remain behind in the midst of higher cultures, as fossilized practices and ideas. Such practices and ideas have, for some reason, not developed with the progress around them. Through them we can, as it were, look back into human understanding as it used to be in prehistoric times.[8] The idea that animism is a prehistoric, ancient belief system, is still around, as well in the more racist versions of those who hold it to be backward and undeveloped, as well as in the romantic, 'hippie' versions that see the belief in spirited nature as more original, pure, not spoiled by civilization. Both positions deny the contemporaneity of peoples who live in a spirited world with the moderns. And both deny the possibility that their ways to be in the world are the product of a conscious choice.

Harvey and likeminded thinkers, whom he calls 'new animists,' avoid such chronological hierarchizing and adopt a non-dualistic relation to nature, which they consider to be neither pure and original, nor primitive and backward. On the contrary, they take it as necessary to secure our future relations with our fellow-beings with whom we share this world. They hold 'that the world is full of persons, only some of whom are human, and that life is always lived in relationship with others' (Harvey 2017 [2005]: p. xvii). The new animists, despite their decolonial motivation, are not primarily interested in culture as a carrier of difference, nor in the historical and political events that created the othering of indigenous peoples in the first place. Their primary interest is ethical, which makes them look forward, to better relations with nature, which makes the view backwards, to understand how those relations came to be spoiled, mostly absent from their work. Australian philosopher Val Plumwood, who called attention to the necessity to rethink our relation to nature from the 1970s on, called for decolonization toward indigenous peoples as well as nature on a concrete, activist level, for instance by giving places their pre-colonial names back. She also worked together with the Aboriginals to restore our (human) relations to nature, and to overcome the divide created between 'them and us.'

Deborah Rose Bird wrote an insightful article on Plumwood's life and work after she died. In it she remarked:

The two key points – recognition of personhood beyond humans, and centrality of relationships – are integral both to Indigenous animism and to the kind of rethinking Val was calling for.

(Rose 2013: p. 96)

What is left unreflected here is the question what it means if moderns adopt an animist position, as self-proclaimed allies of indigenous peoples, without first asking them if and how such solidarity is welcomed. The risk if moderns take the initiative in this, is that unconsciously they will reproduce the politics of epistemology at work in post- and neo-colonial power relations. Their well-intentioned movement may ignore what is really at stake in indigenous relations to nature in the postcolonial context, and thus prolong the silencing of indigenous voices in philosophy. What the new animists promote is a less-modern way to be in the world, which is urged along by the scary conditions our earth environment seems to be in after industrialization. Their position seems to be those of moderns looking for support or help from the indigenous peoples:

Val understood that Aboriginal Australians always live within a world that is buzzing with multitudes of sentient beings, only a very few of whom are human. She thought that a good way to start a major cultural rethink would be to talk with people who are now living within the kinds of understanding we are seeking.

(Rose 2013: p. 95)

The point of departure of this 'rethink' is the present condition in which modernity has brought the world – a condition of environmental disasters, quickly disappearing forests, a declining variety of animal and plant species and a continuously growing pressure of modern consumption styles on the lands of indigenous peoples. The point of a rethink is, that experiences of nature, experiences of being related with non-human-animals and non-animal other (living) beings, are being put to work for an ethical goal. In this case the goal of transforming modern culture to a more sustainable version of itself, hospitable to others, acting for their good. As laudable as such a move may be, it is not what we are after in this book. The work which, to my view, still has to be done first is to accept in principle the end of the position of power of the moderns. Otherwise modern culture will not stop reproducing, even under the flag of morality, existing relations of inequality.

Another aspect of the new animists' work in which the neglect of an analysis of power systems can be detected, is their bypassing the question of ontology. They tend to see the question of what is real as a diversion from the ethical and political work to be done. Moreover, they see ontology as we in the West understand it, as a modernist concept that doesn't support the ecological activism that they are looking for. Once more in the words of Rose on Plumwood:

> ... she was not making a set of truth claims about the world, but rather was asking what kind of stance a human can take that will open her to a responsive engagement in relation to nonhuman others.
>
> (Rose 2013: p. 97)

When Harvey does use the word ontology, he makes clear that in indigenous ways to be in the world, it is not about what is real, but about how to relate to others.

> The force of Anishinaabe ontology ... is that being a person is all about relating. Personhood is not a state but a process. It is not something possessed but performed. To be a person is to act with, among and towards others. It is a 'becoming-with.'
>
> (Harvey 2017 [2005]: p. 227)

Even though this interpretation of an understanding of reality in terms of a dynamics of becoming may make sense for a new understanding of our relations to nature, we need preliminary negotiations before 'we' (the moderns) are in the right position to interpret the words of indigenous peoples' 'communication officers' – those individuals who have taken it upon themselves, like Davi Kopenawa, to be ambassadors to the non-Yanomami, the non-Lakota, the non-Warlpiri (etc.) world. Passing over the question of ontology too quickly could be a remnant of the othering that the animists aim to overcome. It passes the possibility by that Western ontology is just a variety of something more widely dispersed – in the sense that we can think of the reality to which Lakota people are committed, the reality to which Frankish people are committed, etc. This would have for a consequence that any such ontology that 'goes global' – offering an enrichment to understanding reality for all others, will have to submit itself to a global critique, as any one of them imposing itself upon the others would 'forget' what the localized conditions from which it sprang in the first place. In Chapter 6 I will delve deeper in options to understand ontology in the situation of interculturality in which we (humans) now increasingly find ourselves. That investigation will also help to enlighten some points in the so-called ontological turn in anthropology to which I will turn in the following paragraph, a turn to indigenous thinking that came about simultaneously to the new animism, but from a different goal and perspective.

Speaking trees and telling animals

A central idea in animism, be it old or new, is that all beings can communicate with each other, as all beings are or have spirit – or, to say it differently: all exist in infinite communicative, energetic relations which can be more or less active, and take multiple modes of transfer. What spirit is, is an immense question in itself and cannot be treated in a definitive manner in this book. Even if we cannot decide on this question, we may still speak of spirit, and of spirits, in a *cross-cultural* and even *transcultural* manner,[9] to translate the many different ways to understand and to name that aspect of reality that concerns the soul, the inner life, the energetic realm or whatever culturally conditioned concepts we may have – concepts that are bound by practical frameworks to help humans deal with the world. In cultures where nature is considered to be spirited, and communication with non-human beings is held to be indispensable for human life, trees are among the beings that hold a special symbolical place in human life. Trees have a central place in many religious and mythological narratives. Buddha sat under a tree when he had his decisive revelation of the meaningfulness of all life. The tree of life is a central symbol in Judaic, Islamic and Christian traditions and beyond them as well. Trees are also markers of important social structures, such as the palaver tree (cf. Chapter 8, below) that marks the place where justice is spoken in many African societies. Trees were used as 'history books' in African and Australasian cultures in the past (carving markers of important events in their bark), and as archives that anchor lovers' bonds. Apart from those symbolical functions, trees have always provided shelter, medicine, food and drink to humans, in the form of branches, roots, barks, fruits and leaves. These provisions as such may again be understood in spirited ways, as gifts of the trees, or of spirits that are connected with or living within the bodies of those trees.

In social science approaches, as well as in environmental humanities, the above-mentioned 'meanings' given to trees, be they religious or rather embedded in 'folk' beliefs, are generally analyzed in terms of symbols that embody relations between humans – their social practices and institutions. Such analyses, especially when they are informed by recent political, economic and cultural developments, produce helpful descriptions that shed light on how human relations to nature are continuously shifting in the light of present needs to negotiate their environment. A good example of innovative investigations of this type can be found in a recent book on 'sacred groves' in Africa. Its approach is clear from a statement in the introductory article by editor Michael J. Sheridan:

> Sacred groves were symbols of power, locality, ownership, and authenticity on the 'internal African frontier,' but the far-reaching changes of European colonization and the construction of postcolonial nation-states shifted the locus of power into new institutions.... As icons of autochthony, sacred groves often lie squarely at the center of ... struggles over categories and meanings, and are thus concentrated forms of symbolic capital.
>
> (Sheridan 2008: pp. 24–5)

Here power is understood as a real relationship between humans, expressing the scope and limits of agency of individuals or groups over against each other. Ever since Hegel's analysis of the slave-master relationship, the social sciences work with a concept of power as a relation that is both ambiguous and fragile, even when its realization can just as well be brutal and inescapable by those on the receiving end.[10] Analyzing natural environments that, in the words of the new animists, consist of communities of non-human (or: more-than-human) others, solely in terms of their function in inter-human power relationships, excludes the possibility that these communities *themselves* have entered into meaningful relations with human beings. In the case of sacred groves these are communities of plants, birds, fungi, minerals, worms, insects, mammals, snakes and others, who live interconnectedly in networks that are to the human eye most typified by trees. Sacred groves appear as (little) forests, which humans can enter to seek not only nourishment or shade, but first and foremost communication with non-humans, spirits, be they the spirits of those living others, or spirits dwelling there. Animist approaches, in contrast to more secular ones, take into account that power is not only to be negotiated between humans, but between humans and all such others as well. Even the earth itself is active, and 'speaks,' at least according to Manngaiyarri, an Aboriginal thinker:

> That's what this earth makes you to do. Makes you go this way. Or you go up here. You get up first thing in the morning, when you camp you get up first thing in the morning, and you go. That's the word earth gives you – whatever way you go, see?
>
> (Jimmy Manngaiyarri, cited in Rose 2013: p. 105)

An intercultural approach has to take these multiple relations of humans with what is around them into account, to be able to secure sustainable conditions for dialogue.

The question which I will look into now is what representatives of the ontological turn in cultural anthropology may have contributed to such interculturality. One, of them, Eduardo Kohn, especially focuses on the forest as the place of animist understanding of the world. In his book *How Forests Think* the forest is chosen as the descriptor of the 'animist' way to live in the world of the Runa of Avila, a community in the Amazon, that is

> ...far removed from any image of a pristine or wild Amazon. Their world – their very being – is thoroughly informed by a long and layered colonial history. And today their village is just a few kilometers from the growing, bustling colonist town of Loreto and the expanding network of roads that connects this town with increasing efficiency to the rest of Ecuador.
>
> (Kohn 2013: p. 3)

Despite their closeness to the modern world, the Runa relate to nature as to communities of trees, dogs, jaguars and other living beings, which communicate to them.

In *How Forests Think* (2013), Kohn describes, using the sign theory of American pragmatist J. S. Peirce, how the Runa language reflects the agency, the willing and thinking, of the non-human others among and with which they live. Peirce's semiotics which takes signs to '... stand for something in relation to a "somebody"...' (Kohn 2013: p. 75), allows Kohn to assign selfhood to every being that gives off signs, that signals. Describing how, for instance, hunters for birds seek out fruiting trees that attract those birds, he shows how the perspective of the birds can be known by the humans. Analogously, one can say that the tree's perspective can be read from the fruits, which signal its intent to spread its seeds around. Thus, without having to go too deep into spirit ontology, Kohn can clarify to Western readers how the Runa live in a world that is full of all kinds of selves. The selves Kohn recognizes, can be other than human, but must be biologically living entities. Kohn departs here from what the Runa themselves hold to be, and limits himself to his own constructed 'sylvan' ontology of life. The Runa themselves think in more complex ways about being spirited, attributing selfhood to 'non-living' entities as well as to non-embodied spirits, such as demons. This diversion from the thought of the people he studies, highlights that Kohn's ontological turn involves a process of reconstruction that does not necessarily aim at a 'true' representation of 'animist' ways of being in the world. In Kohn's words:

> I recognize of course that those we call animists may well attribute animacy to all sorts of entities, such as stones, that I would not ... consider living selves. If I were building an argument from within a particular animistic worldview, if I were routing all my argumentations through what, say the Runa think, say or do, this discrepancy might be a problem. But I don't.
>
> (Kohn 2013: p. 94)

His reason for diverting here from Runa ontology, can be found in his desire to make general claims about the world. Thus what started as a transformed kind of cultural anthropology, therefore turns into a general philosophy that aims to 'go global.'

Here we see what may happen when cultural anthropology doesn't make use of the methods developed by intercultural philosophy, and bypasses a decolonial analysis that includes the postcolonial power relations that prescribe what kind of epistemology is accepted as valid and what not. Doing general philosophy, according to Kohn, would not be possible starting out from the world in which the Runa live. He adds that he even doesn't want to find out how this people thinks about forests, but how forests think. A decolonial analysis would understand both epistemic worlds – that of the thought of forests and that of the thought of a certain people, caught in modernization processes they have not sought for – to be interrelated, and their suppression as well. Overlooking the epistemological aspects of the situation in which the Runa people live, as Kohn does, may have for a consequence that we overlook their actual resistance to the generalizing theoretical hegemony that undergirds their colonial suppression and exploitation. In that sense another generalizing reading of elements of their ontology must remain the newest expression of the modern Western epistemological hegemony. This then obfuscates the question as to how and where the environment (theirs, ours) can be negotiated. It means another appropriation of an indigenous ontology, a passing by of the possibility that the 'weird' talk of demons and spirits, next to the relating to trees and animals as selves, conveys a relation to nature that in a truly postcolonial condition should be heard just as well as the reconstructed ontologies of the anthropologists.

The talk of animals is, for peoples like the Runa for whom hunting is important, the most obvious set of signs. Kohn words it as follows:

> Runa animism grows out of a need to interact with semiotic selves in all their diversity. It is grounded in an ontological fact: there exist other kinds of thinking selves beyond the human.
>
> (Kohn 2013: p. 94)

The jaguar as a self sees all kinds of animals as prey, among them humans. The human needs to be aware of the gaze of the jaguar to save himself. Recognizing animals to be selves creates a moral problem for the hunter: eating them may be considered as a form of cannibalism. In this context anthropologists and new animists alike have tried to understand ways to be in the world that center animal–human relations in terms of 'metaphysics of predation' (Viveiros de Castro 2014 [2009]: p. 144). Viveiros de Castro theorizes that Amazonian peoples, of whom some in the past (the seventeenth century, Viveiros de Castro 2014 [2009]: p. 140) may have practiced anthropophagism (cannibalism),[11] understand killing and eating others as the constitution of the self. Eating the other means incorporating the other to be the basis of being a self, so that they perceive the other in the self. His theorizing follows the (post)structuralist work of Lévi-Strauss, who became most famous by his 1962 work *The Savage Mind*. (Post)structuralism deduces meanings from seemingly incomprehensible indigenous categorizations, by discovering patterns in them, deep structures, that express the normative-ontological orderings of non-Western cosmologies. An example from Viveiros de Castro:

> …what was really eaten in this enemy? … The 'thing' eaten, then, could not be a 'thing' if it were at the same time – and this is essential – a body…. What was eaten was the enemy's relation to those who

consumed him; in other words, *his condition as enemy*. In other words, what was assimilated from the victim was the sign of his alterity, the aim being to reach his alterity as point of view of the Self.

(Viveiros de Castro 2014 [2009]: p. 142)

Here, as in Kohn, we see the echoes of structuralism and philosophy of language, with its attention to semiotics, pragmatics and grammatological deep structures in cultural practices. Although such an approach seems at first sight to take the reality of indigenous peoples more seriously than more descriptive anthropological work that remains within the bounds of Western perception, it doesn't practice *dialogue*, or even realize the need for a *negotiation* that should precede it. The politics of epistemology that dominated anthropology from its beginnings is, alas, still at work even in these critical writers that claim to think along postcolonial lines. Viveiros de Castro seems to acknowledge so much, however, when he admits that '...anthropology, even if colonialism was one of its historical a prioris, is today nearing the end of its karmic cycle...' (Viveiros de Castro 2014 [2009]: p. 40). He even criticizes its postcolonial versions that claim that indigenous peoples have only been represented as other in anthropology, and never have come to word. Viveiros de Castro hopes to bring the destruction of anthropology to its end by showing

> ...that every nontrivial anthropological theory is a *version* of an indigenous practice of knowledge, all such theories being situatable in strict structural continuity with the intellectual pragmatics of the collectives that have historically occupied the position of object in the discipline's gaze.
>
> (Viveiros de Castro 2014 [2009]: p. 42)

His argument moves too fast to produce a true mending of the othering of traditional anthropology. To just ascertain that in anthropology, indigenous peoples have always already had the word, even while being object, denies the distortions made in the descriptions of their ways of life – the projections of purity and wilderness, or of primitivity and backwardness. However we value Viveiros's account, his solution ignores the importance of criticism, which is, as argued above, the main instrument philosophy can bring to the struggles, if not the wars, in which indigenous peoples unwillingly have been drawn (the wars over resources on and within their lands, struggles in which their cultures and even their world, and our commonly shared earth as well) is at stake. In the end, the ontological turn in Viveiros de Castro, as in Kohn, has all the characteristics of a theoretical appropriation of indigenous ways of being in the world. It doesn't dialogue, it doesn't try to let the indigenous peoples themselves speak, and it reduces their complex spirit ontologies to functionalistic analyses of material life (eating and being eaten).

The materialistic approach arose as an alternative to that type of early ethnological work that read the said ontologies from an interpretive frame that was heavily colored by Christian theology. In that kind of work, indigenous experiences of the spiritual realm would be described as 'natural' and 'primitive,' and opposed to the 'higher' 'civilized' ones of 'true' religion. To avoid such misconceptions, many post-1960 authors have taken resort to functionalistic and structuralist interpretations of the spiritual. The categories etic and emic were introduced to free the Western researcher from any commitment to the spirit ontologies they studied. In the realities of our present postcolonial world, however, trading a condescending approach for one that doesn't hear the spirited language in which indigenous peoples express their way to be in the world, is not acceptable. In this age, research without a deep commitment to listen to the other, to be critical toward one's own tradition and to be open to true negotiation, is a waste of time (cf. Kopenawa and Albert 2013 [2010]: pp. 5–6).

Shamans and spirits

The approach taken in this work differs from the 'new animism' as well as the ontological turn in anthropology in several aspects. The present work does not aim at a 'hermeneutics of identity' (that is one that seeks a fusion of horizons), which would mean describing animism as a clear metaphysical system that is extended across the divide created by colonialism and its consequences. I do hold it to be possible to arrive at transculturally shared human relations to nature. To be able to do so presupposes actual cross-cultural investigations, and taking the consequences into account that the conflicts between modern and indigenous ways of being in the world have for our articulation of human relations to nature. This also means that we should deconstruct the articulations expressive of power systems that other the indigenous peoples, in romantic or dehumanizing descriptions of their way of life. As a consequence, it is too early to center an ethical focus, as ethics as we understand it since the European Enlightenment (as knowing what to do, independently from what religion or nature tell us – cf. Roothaan 2005) may be part of the problem. What has to come first is an investigation of the possibility that 'spirit ontologies' or shamanistic ontologies, are true, and may critique white and modern ontologies.

Dealing with spirits and some form of shamanism in the broadest sense always go together. In the first chapter I explained two aspects that are to be noted in my use of the term 'shamanism' and 'shamanistic,' as in shamanistic ontologies. The first aspect, concerning shamanism as a practice in dealing with spiritual beings, entails that I will use this concept to refer not only to practices of tungus shamans, as described by anthropologists, nor only to practices of more or less traditional living indigenous peoples, but as mediatic practices that are available to human beings across cultures. Harvey has also noticed the loose relation between traditional living styles and shamanistic practices, where he writes that

> ... there is no absolute correlation between the existence of shamans and subsistence by hunting – or hunting and gathering. There are shamans in pastoral societies (as there are among urban ones) and they might engage with animal persons or their 'owners.'
> (Harvey 2017 [2005]: pp. 146–7)

All the same, shamanic practices and/or a shamanistic understanding of the world, are identifying elements of those peoples that are commonly referred to, and that refer to themselves as *indigenous*. They mark a distinctive feature of how these peoples organize their societal relations as well as their relations to the larger world. What is distinctive here is that shamanism is recognized as indispensable for human life, providing practices of healing, of reconciling, of keeping cosmological categories in place, and, last but not least, of finding direction for individuals as well as communities in moments of choice. When people modernize, they start to reject such practices, following the dominant modern discourses of science, religion and philosophy, and start to see them as superfluous ritual, providing irrational, superstitious or non-valid knowledge. The second aspect is that in shamanistic cultures ontologies are shared among shamans proper and other members of society, that allow shamanic practices to be recognized as procuring valid knowledge and therefore to have generally acknowledged pragmatic reality.

So we have shamanic practices, that are cross-culturally available, and shamanistic cultures, in which the validity of those practices is generally accepted, in connection with shamanistic or spirit ontologies. I describe the relation between ontologies and practices slightly differently from the way Harvey does. In *Animism* he points out that shamanism is not a specific religion.

> The fact that shamans shamanise does not make them, or the groups for whom they shamanise, members of a religion called Shamanism.
> (Harvey 2017 [2005]: p. 137)

In so far, if we accept that 'religion' may be substituted for 'culture,' this does not conflict with the distinction I made here between cultures and practices. The next step Harvey takes brings him on a different path, however:

> Their religions are animisms not shamanisms ... [because of] the understanding that shamans and shamanising provide particularly powerful tests of the boundaries of human attempts to find appropriate ways to live alongside other persons, i.e. of animism.
> (Harvey 2017 [2005]: pp. 137–8)

The problem with this step is not that it distinguishes 'shamanizing' practices and religion (or culture), but that it suggests that we may define animism as a clear-cut metaphysical understanding of our being in the world as a being in relation as persons with other persons, most of which are non-human. As I already pointed out, shamanistic cultures will vary considerably in their understanding of relations of humans to 'others,' which is an argument against using 'animism' as a term indicating a specific type of religion (culture). Further, I prefer to speak of relating to spirits rather than to persons – for two reasons. The first one is that it demands of us to get clarity about the ways we can understand the category of spirits first (and not reduce it prematurely to the Western concept of personhood); the second one is that it preserves the potential spiritual understanding of human being in the world, before reducing it to material processes such as eating and hunting, as we saw happening in the above quoted texts by new animists as well as representatives of the ontological turn.

My insistence on characterizing indigenous cultures by the concept of shamanism rather than that of animism rests, moreover, on the fact that my characterization puts epistemological questions (what and how can we know?) central, in contrast to the characterization as animism, which passes the epistemological question by. Adhering to some form of animism, as in the new animist movement, does not necessarily imply the recognition of shamanistic knowledge as valid knowledge. Researchers of religion, environmental ethicists and anthropologists that are most sympathetic to indigenous ways to be in the world may recognize the effective reality of spiritual presences, while still ignoring the epistemic discourse that is dominant in the cultures with which they are in contact. They may take recourse to complex theoretical constructions (like Viveiros de Castro and Kohn) to include shamanistic content in their work, meanwhile reducing it to material processes (hunting, eating – which both mentioned authors do, as well as Harvey and Plumwood) or treat it as some kind of mysterious 'wisdom.' A description of shamanic practices as wisdom can be found in the work of anthropologist Paul Stoller, who practiced participatory observation among the Songhay in Niger. To further clarify why I insist on recognizing the importance of epistemological questions around shamanistic ontologies, and the connected issue of the politics of epistemology, I will discuss his work in some more detail here.

Stoller's work is of the next generation of anthropologists after the structuralism of Lévi-Strauss and his contemporaries. Following the work of Clifford Geertz, Stoller's phenomenological investigations aim to provide 'thick descriptions' of the lives of his research subjects. From Derrida and other poststructuralists he borrows the insight that the search for objective truth is conspicuous and may be connected to totalitarianism in politics (Stoller 2004, pp. 199–200). To avoid such a connection Stoller practices a narrative style, which aims to uncover the pragmatic reality of things. The descriptions he thus produced of what he calls 'possession' or 'sorcery' among the Songhay in Niger (Stoller 1989) are impressive in their intimacy, while he weaves complex descriptions through them of the cosmological/spiritual hierarchies and the family structures which organize Songhay spiritual mediation. The wide array of knowledge practices Stoller describes, from a first person perspective as well (having been a pupil of the 'zima' [village priest] Adamu Jenitongo), can safely be labeled 'shamanistic' in the way I use the concept here. Varying from spiritual warfare with opponents, through herbal knowledge to cure people, to mediating the chaos outside the village to protect the order of community life within (Stoller 2004: pp. 134–5 and 186), these practices all belong to the array of what shamans do. This goes as well for the geomancy Stoller learns from his master, as geomancy is practiced in some form or another by many shamans worldwide.[12]

Shamanic practices are not reserved to shamanistic cultures. People can do shamanic things in whatever cultural conditions. Shamanistic cultures, however, are characterized by the fact that they recognize shamanic knowledge as valid, and foster and sustain techniques, teaching institutions and a general worldview that help to keep shamanic traditions alive. Also, they recognize the special and much respected role of an initiated shaman, who has learned all kinds of shamanizing techniques from a predecessor. As interesting as Stoller's description of the spiritual life of the Songhay in his 1989 book is, for our purpose his 2004 work on sickness and healing is more interesting. In this book – *Stranger in the Village of the Sick* (2004) – he returns in his memory to the Songhay, and especially to the lessons of Adamu Jenitongo, when he has to deal with cancer, while living once more in his home country, the US. He narrates how on many occasions the memories of his time in Niger come back to him, to help him deal with the illness, and, the other way around, how his dealing with illness helps him to more fully understand what Jenitongo taught him. He narrates how, on his first return to the US, he kept to certain practices he learned in Niger:

> When I returned to my 'Western' life, I tried to maintain a tie to the ways of sorcery. This link comforted me and reaffirmed my connection to Adamu Jenitongo. I made a small altar in my home. It is a low round table covered with a black tunic that represents Dongo, deity of thunder. Positioned at the center of the table surface is a *batta*, a sacrificial container.
>
> (Stoller 2004: pp. 94–5)

He describes how he practices ritual cleansing, recitations, song and divination – also to help family and friends to regain 'health, harmony and wellbeing.' During the time he had to undergo chemotherapy, he finds such a way of life helpful to keep '... a certain sense of personal control, which goes a long way toward maintaining quality of life' (Stoller 2004: p. 100). Also his practice of rituals helps him to see things more clearly. His potential for vision was given to him by his mentor way back in Niger, with the warning that the potential can only actualize at the right time, and nobody knows when that will be. Things Stoller learns are patience, waiting for that right time, as well as realizing the modest place an individual takes in the collectivity in which he lives, including past and future generations. We cannot reach any form of absolute insight, or complete fulfillment in life – the important thing is to realize you have to learn what is there to learn for you, and then to pass it on to those around you and who come after you.

As an anthropologist, Stoller now, looking back, finds that explanation nor description are his task:

> I now believe that the anthropologist's fundamental obligation is to bear witness.... In the end this turn may take us to that elusive and oft forgotten end of scholarship: wisdom, the knowledge that enables us to live well in the world.
>
> (Stoller 2004: p. 200)

When reading this conclusion of his book, I was a bit disappointed. The modesty of Stoller's version of anthropology is appealing, as it avoids the kind of power grab that ethnological explanations of cultural life have often shown. What I had expected was, however, a more in depth *discussion* between the wisdom he learned from the Songhay and the American culture of which he is a member. A discussion on what shamanic practices convey about human life in the world, and our relation to nature. It is precisely the modesty taken up after the more ambitious attempts at explanation and description of earlier anthropologists, that keep Stoller from going there and make him halt before entering the discipline of philosophy, which would (and should) attempt such a discussion.

In this chapter I discussed works from different disciplines, such as environmental ethics, religious studies and cultural anthropology, and investigated what they might bring to a philosophical critique of the othering of indigenous peoples, their knowledge and their shamanic practices – the othering that modern Western culture produces to define itself. All of them provided relevant material to critique the existing power balance, and expressed attempts to relate to indigenous peoples in more respectful ways. Among those is the realization that shamanic work should not be psychologized, to be found in Harvey:

> ...academia has maintained an 'exotic other' to strengthen the perception of its (or Western) superiority ... The psychologisation of interest in shamans is part of that colonizing process.
>
> (Harvey 2017 [2005]: p. 142)

Our challenge now goes beyond such attempts at understanding or describing shamanistic cultures respectfully, and will take the possibility serious that 'spirit idiom' is true (Ellis and Ter Haar 2004: p. 6), refers to reality, and that shamanic practices make it possible to relate to spirit(s) for real, expressive of a certain relation to nature that differs from the modernist one. Such a relation to nature critiques modernist dualisms, showing these to obfuscate realities we need to survive and live well on this earth. The fact that taking shamanistic knowledge seriously is so uncommon in philosophy, stems from the formation of its dominant Western variety, which meant the banning of the spirits, as well as of ontologies that recognize their reality. We will turn to this process in the following chapter.

Notes

1 This makes it clear that many internal criticisms of modernity are inspired by Freudian ideas. Freud, to be sure, dethroned the idea that 'man' was all reason and goodness, and ascribed to him gender – femininity or masculinity, bodily drives and emotional disturbances, all influencing thought and behavior. Together with Nietzsche, Feuerbach and other thinkers who argued to think of the modern subject as embodied, he helped to undermine man as identical with himself and divided the human being within itself.

2 Randall stresses that as a Warlpiri man he does not participate in the place as the local Arrernte would – '[...] he does not have the rights and responsibilities given by being consubstantial with the place' (Harvey 2017 [2005]: p. 76).

3 Initiated by authors such as Heinz Kimmerle, Ram Adhar Mall and Franz Wimmer.

4 In the late 1960s the initial interpretation of James's work as truly American and not connected to European philosophy was being turned around (cf. Edie 1970). Since then we see a growing body of work that investigates the connections of his work to developments in European psychology and phenomenology (cf. Taylor 1996).

5 We see this as well in a late version of an influential phenomenological approach, that of Charles Taylor, who in his *Sources of the Self* aims to recapture and articulate what modernity in all its phenomenological richness was essentially about (cf. Taylor 1989).

6 To highlight the differences in the articulation of what it is to be human in modern, postmodern and non-modern philosophical discourses, I intend not to 'correct' throughout this work, the non-inclusive, non-diverse language of classical modern philosophy. Wherever I do write in an inclusive style, it is clear I have moved to a place where modernism is deconstructed and criticized.

7 A more elaborate argument of the consequences of Kantian ethics for the relation of human beings to nature is to be found in Roothaan 2005.

8 We may see here the echo of the debates on fossils that were so central around the time that evolution was first introduced into biology – the evolutionists saw in them a proof of the development of organic life forms, whereas those who held to the Bible explained them away by supposing God created them as fossils, together with the life forms we know since biblical times.

9 An analysis of the meaning and necessity of these approaches is given in Roothaan 2019b.

10 As fatal as power may be, it still rests on (suppressed) freedoms of thought, of speech and of action, and remains vulnerable to criticism and resistance, as the

last resort of power, to kill, can only be applied within limits. If the oppressor would kill all those in his power, he kills his power itself.

11 Harvey however criticizes taking literally historical accounts of cannibalism. See Harvey 2017 [2005]: pp. 151 and further.

12 Whether one observes patterns of kauri shells, the footprints of wild animals or even 'tarot cards' – all such practices are shamanistic in the sense that they aim to read knowledge that is somehow stored in bodies. In the case of tarot or I Tjing, it is the hand that draws or throws that releases information. Wild animals can be understood to release information that concerns human problems as their bodies are better tuned in to energies that define the situation. All forms of geomancy make use of a system of interpretation, comparable in structure to modern scientific tables to categorize elements in nature.

Chapter 3

When the spirits were banned
Kant versus Swedenborg

Moral and epistemic ambiguities

To understand present day Western relations to nature, we have to go back to seventeenth and eighteenth century Europe when the modern scientific approach gained ground. Especially significant for our aim – to understand these relations in light of the question whether nature is taken to be spirited, and spirits to be taken into account to live good lives, or whether their acceptance is seen as ignorance and nature itself as disenchanted, devoid of spirit and from this perspective passive or dead. In the center of the crossroads, when the modern view of nature gained final public acceptance, stand two important writers – one defending (within the bounds of modernity) a completely spiritualized experience of life, the Swedish mining engineer and spiritual writer Emanuel Swedenborg, the other constructing the boundaries of pure reason and excluding spirit experiences from the publicly acceptable trustworthy knowledge, the philosopher Immanuel Kant. In this chapter I will first describe the precursors of the issue of the (non-)acceptability of spirit ontologies in the seventeenth century. Next to the most famous epistemological argument against spirit ontologies, the ethical argument was central. After discussing René Descartes's systematic doubt of all human knowledge and pastor Bekker's ethical critique of spirit belief, we will turn to why and how Kant aimed to show Swedenborg was wrong, and to the content of Swedenborg's modern version of a spirited understanding of human life and the world.

Because of the success of the novel sciences, their methods soon became the paradigm for trustworthy knowledge. A combination of experimentation and mathematics became the norm to understand nature, and philosophers tried to emulate this approach. Providing an organized manner to produce very precise results, experimentation made everyday experience appear as an imperfect, chaotic and untrustworthy source of information. Newly designed instruments were used, that were considered to offer more precise information than the human senses. The new approach to nature was founded on the objectification of anything that can be perceived. Famous for articulating this foundation is René Descartes (1596–1650), who said that all being is either mind or matter: a knowing and thinking subject, or a passive object. The latter understood to be moving, no doubt, and taking its place in causal chains, but doing so following purely mechanical rules. This divide makes the conscious thing, the reflective mind, that can say 'I doubt,' 'I know,' 'I am,' more real than the non-thinking object, for in its reflection, doubt and will show a certain freedom that all non-human being lacks. At the background of this dual world is God, now seen as a divine Watchmaker, guaranteeing the congruency of what pertains in the realms of mind and of matter.

These changes had massive consequences for the relation to living nature, as the biological realm now was more and more understood through a mechanistic lens as well, just like the physical. This meant that all that is biologically living, *ontologically* becomes part of passive, *dead* nature. This denial of a conscious spirit to all living nature is part of the disenchantment that became dominant in philosophy. Although this doesn't mean that nobody took a spirited view of nature serious anymore, it signals the final end in Europe of shamanic practices as public culture. Conceptually declaring nature to be non-human, devoid of persons and thus morally 'dead,' makes room for invasive scientific investigations of non-human life. French thinker Lamettrie thinks it is acceptable to dissect and investigate animals without any remorse, as they cannot truly feel pain. The scientist who cuts open a live animal to see how the heart works, not only does not need to fear any potential moral or spiritual harm coming back at him anymore, but does not even need to feel moral pain. Dutch philosopher Spinoza thinks that eating animals is completely morally neutral. All in all, we are now far removed from the shamanistic worry that

> [...] human food consists entirely of souls [...] like we have, souls that do not perish with the body and which must therefore be propitiated lest they should avenge themselves on us for taking away their bodies.
> (Harvey 2017 [2005]: p. 144)[1]

The modern approach to nature frees human agents from fear of other beings with souls, those with bodies (animals, plants) as well as those without bodies (free spirits or the spirits of deceased). The psychological effect of this change can hardly be overestimated – the earth now becomes a free playground for mankind, where he can roam freely, investigate without fear, dig up and transform resources, cut open plants and animals, and use whatever makes his own life more secure and comfortable. The only fears left to men now concern either that the (morally and spiritually neutral) forces of nature may cause natural disaster, disease and natural death and the aggression of other men.

A fundamental problem with this new approach to nature, as conceptualized by Enlightenment philosopher Immanuel Kant (1724–1804), is that a moral capacity or spiritual agency is now not only absent from living beings such as trees or animals, but that in fact it can neither be located in the human being – except for its rational capacity, man is just an animal, with desires and dislikes, a living machine possibly even. Ethical knowledge to Kant should be autonomous, self-legitimized and not based in the animal nature of human beings, for that animal nature is passive in the moral and spiritual sense – it cannot know, it cannot will. This awareness creates a moral problem for Kant – how can he explain that human beings make moral choices, how can they consider themselves to be *free* to do so? To solve this problem, Kant proposes there to be two realms: the empirical and the intelligible realm. In the first one, all is known in its objective, passive being – be it human or non-human. In the second one, the rational will is active, that can choose what it can know to be good. In this realm, we (humans), inasmuch as we are rational beings, are detached from our empirical, animal nature and can detect what is unequivocally and universally good, bypassing all those parochial goods that come with natural alliances to family, clan, the village or the land on which we live. The modern moral agent that Kant helps to cut out thus floats free, is a citizen of the cosmopolis and the keeper of the rules of what we now have named global justice. This justice does not reckon with any local or embodied loyalties, and may inform any actions that choose the good for the global community of humankind over local, particular communities.

As progressive as Kantian ethics was with respect to old feudal relationship structures, it created serious moral ambiguities as a consequence, as actual, living nature, and the people living in direct contact with it, were now deemed irrelevant from a moral point of view. Kantian ethics, even in all the adjusted varieties that still dominate Western ethical thought, is an ethics for modern nation states, for international legal and moral bodies, for all the bureaucracies that keep these soulless bodies alive, as well as for the spiritually detached citizens that live in them. It discriminates against indigenous peoples, who are now considered stateless and therefore a threat to European modernity. They are no citizens, they are categorized as 'wild,' and do not count from a cosmopolitan perspective. The consequences of this universalizing approach can be seen even today, for instance in the workings of global wildlife protection agencies. While they work to preserve lions, monkeys or elephants for mankind (morally good from a non-local perspective) – they sometimes force communities of traditional semi-nomadic hunter-gatherers off protected lands, or into a sedentary life without any political autonomy, and no right to relate to the nature around them according to their own traditional knowledge and customs. They may be given jobs as game keepers in the best case, and in the worst case they can be persecuted for keeping to their traditional ways to feed themselves through hunting. What goes for them, goes analogically for the animal and plant life that is increasingly being brought under the rules of the conservational parks. Animals have to learn to stay in those parks or they will still be shot, for fear that they will endanger farmlands or humans whose habitat has encroached upon their territories. Conservationists study nature and decide what species of plants and animals will be protected, and will make interventions into natural habitats to accommodate these species, while denying freedom and agency to the peoples who will not fit the image of the man in possession of universal reason, and the principles of global justice.

The core of this kind of moral ambiguity consists in the idea, that when all that is natural in us is non-rational and therefore non-conscious, devoid of knowledge of the universal good and without moral will, humans as 'empirical selves' cannot be moral. Our 'gut feelings' concerning what is good are therefore not really moral, and have to be corrected by what we know to be good through our moral will. Also any 'higher' moral feelings, as they come to us for instance in religious experiences, in conversion experiences and in mystical insights in our connectedness to all creatures, are not moral in so far as they are not reasonable. True morality, according to Kant, is autonomous, which means free from all natural or religious impulses. This has left ethics after Kant with the problem of how to embody the insights of the rational faculty in real-life practice. Conversely, modern autonomous ethics also leaves all living creatures in so far as they are not reasonable out of the class of moral subjects. When you think of it, the condition of rationality excludes a large part of the human self from moral agency. When we sleep we are not in possession of our rational faculty – nor are we when we dream, or when we are overwhelmed with emotion. The dreamtime knowledge described by Robert Wolff, for instance, is excluded from the ethical realm. When we are subject to psychiatric illnesses, when we are wrapped up in strong moods of depression or of passionate love – even when we love our own kin we do so not out of a true moral feeling. Even people with impaired intellectual capacities cannot be really considered to be moral subjects. Babies can't. Nor people with dementia. Animals (actually, all other living beings) cannot be considered moral subjects. And spiritual beings, if they exist, are only moral in so far as they are rational. In the modern world, therefore, only those spiritual beings that are rational can be agents. All other things, living or not, of spiritual or of embodied constituency, only exist as phenomena, as appearances and apparitions, and cannot tell us anything. They cannot convey any meaning to us, they don't have meaning, they don't have signs or languages. Consequentially all people who do suppose they can, like the Runa described by Kohn, and all the other peoples who consider shamanic knowledge to be possible, have to be considered to err – bringing them into the category of the rationally impaired.

The (dis)enchanted world

In 1691 a Dutch pastor published the first volumes of a book that not only would transform his personal life but also signaled the changing ways in which early modern Europeans viewed spirit ontologies. The work, called *The World Bewitched* (*De betoverde weereld*), was an unexpected bestseller, selling thousands of copies already in the first months of publication. Its author, Balthasar Bekker (1634–1698) set out to challenge the belief in spirits and devils that still prevailed in the Dutch protestant church he served. He was appalled that 'enlightened' Christians, who had forsaken popish superstitions, still practiced magic and adjuration in their daily lives. Although this is not much known nowadays, his work makes clear that many seventieth century Calvinists would call out witchcraft or the devil in case of any adversity. It appalled Bekker, as they should have let themselves be led solely by the Bible and the natural light of reason, as contemporary philosophers and theologians were arguing. What he targeted most passionately were witchcraft trials (which had by then virtually ended in northern Europe) and the denial of moral responsibility that expressed itself in attributing sinful acts to spirit possession.

As things stood, Bekker's views were hotly debated, being far from mainstream. Calvinist orthodoxy (most notably in the person of Gijsbertus Voetius) opposed them, fearing Bekker's denial of the reality of diabolic forces would take away the believers' fear of sin and of evil. As Bekker held that Christians should solely focus on the Goodness and Love of God, his views led him to an anti-confessionalist position, bringing him more and more in conflict with his church. According to church historian Andrew Fix the debate around Bekker's work presented '...the last great debate over spirits in the Dutch Republic' (Fix 1999: p. 5). Actually, Fix notes, although popular belief in the spirit world stuck around in Europe, it was erased in scholarly work, by the scientific worldview that would undergird academia from the seventeenth century on, as well as from the public policies that would base themselves on this worldview. Whereas the protestant ministry was very much focused on academic theology as its foundation, it would later also follow the banning of the spirits that took effect in the European Enlightenment. But not yet in Bekker's days. In fact, Bekker lost his allowance to preach over the controversy that his work stirred up.

The World Bewitched (*De betoverde weereld*; Bekker 1691) is not just a treatise denouncing the burning of people as witches as murder, nor does it only argue for a modernized, more rationalistic (Cartesian) approach to religion. It is arguably one of the first comparative studies into spirit ontologies. Bekker thought he should base his investigation of matters concerning spirits on a worldwide empirical classification of relations with the spirit world. Despite his lack of direct sources, which makes his overview a collection of narratives about distant countries based upon hearsay, he introduced the modern approach of studying spiritual beliefs and spiritual practices separately. After his comparative research, he proposes to critically distinguish sense from nonsense among worldwide systems of dealing with spirits – using the Bible as the spiritual measuring method. The result has been, as we know now, the denial of (evil) spirits holding power over human agency. In order to argue for this position from the Bible, Bekker had to clarify its relevant concepts philosophically – discerning diabolic spirits, human spirits and the Divine Spirit.

His ability to do his investigation, Bekker claims, rests on his lack of fear for any spirit or specter ('schim noch spook'). As spirits are normally seen to have a large effect on the human emotions, instilling fear or, contrarily, peace of mind, the reality of their agency is thought to depend on what humans feel coming from them. This is nowadays often seen as an argument against their existence, including the existence of God as spirit, as the spiritual is understood to be a projection from the human emotions. For Bekker it was still clear that, apart from the human spirit, God exists as the supreme spirit on which everything depends. Modern theistic approaches to reality normally follow this route, declaring nature, the animal kingdom and all organic and non-organic being to be empty of spirit, making humankind 'alone in the universe,' except for their relation to God. This view can be seen as an ultimate consequence of the Abrahamic revolution in religious matters, which stressed the unique bond (through covenant) between God and humanity, first denying divinity of the 'other' gods, much later resulting in the denial of spiritual importance of anything but God and the creature that likens him – man. In its most radical conclusion, in present day secularist science, all spirit is then removed from the world – and the idea of God as well as of the human soul are reduced to the workings of the material brain. Free will, the ornament of human spirit, is declared to be an illusion, produced by firings of neurons, which follow spontaneously, without any previous thought or desire.

Rationalism and empiricism, that are presented in philosophy handbooks as two opposed modern schools of thought, are in effect two sides of a coin, for both deny the reality of the agency of embodied spirit (or quasi-embodied spirit, in the case of ghostly apparitions). Agency is the prerogative – for modern philosophers – of the rational subject (thinking, judging and willing), whereas everything perceived by that same subject is deemed passive, just 'subject matter,' the stuff of which knowledge is formed only actively by the mind. For René Descartes, the philosopher often taken to have first commenced such a line of reflection, criticism of those who take spirits seriously didn't spring from a moral and religious motivation as was the case in Bekker, but from the desire to find an unshakable criterion to decide between valid knowledge and false information coming from either the senses, or from the inner perceptions that take place during sleep or while fantasizing. The effect of this move has surely been that Europeans were increasingly convinced of the trustworthiness of the results of the new natural sciences, while these objectified the empirical, submitting it to rational procedures of measurement and argumentation. In order to convince his readers that empirical knowledge could be trusted when filtered through the criteria of reason, Descartes undertook his famous procedure of systematic doubt. It starts like this:

> I will now shut my eyes, stop my ears, and withdraw all my sense. I will eliminate from my thoughts all images of bodily things, or rather, since this is hardly possible, I will regard all such images as vacuous, false and worthless.
>
> (Descartes 1996: vol. II, p. 24)

In the remainder of his text, this procedure of meditation leads swiftly to the discovery that the criterion of trustworthy knowledge lies in the thinker's doubting faculty. Finally, by means of his argumentation that it is impossible that God would be a deceiver of humankind, this same certainty should convince the reader of the validity of sensory information. Thus the senses were redeemed through an external source, even when they were first considered to be very unreliable in themselves. Of course the other side, that of empiricism, claims the opposite, that no theory and no reflection can be trusted unless it is grounded in empirical fact. Both schools, however, agree on the disjunction of the two. And this again makes the acceptance of embodied spirit or spirited living bodies as unity impossible. Conscious agency and willing now becomes a famous problem in philosophy of mind, as for instance in Wittgenstein and Ryle. When all biological life is dispossessed of spirit, the human mind becomes a ghost in the machine. Spirit appearances now can only be understood as products of this mind, as the conclusion of the *Meditations* reads:

> ...dreams are never linked by memory with all the other actions of life as waking experiences are. If, while I am awake, anyone were suddenly to appear to me and then disappear immediately, as happens in sleep, so that I could not see where he had come from or where he had gone to, it would not be unreasonable for me to judge he was a ghost, or a vision created in my brain...
>
> (Descartes 1996: vol. II, p. 62)

So God to Descartes guarantees empirical knowledge, but leaves apparitions to disorderly workings of our brains. Bekker was a Cartesian, and following him, he could maintain that belief in spirits meant that the human mind had 'bewitched' or 'enchanted' the world, a world which this same mind was capable to *disenchant*. The ingredients for later declarations that those who reckon with spirits are not mentally healthy, dysfunctional or immature (not well developed like a normal human being) are clearly already in place. To conclude, the early modern disenchantment takes two routes – it is defended on grounds of morality (human beings should take full responsibility for their actions and not blame them on spirit possession) and on grounds of epistemology (we can only trust information when it is sifted on the basis of rational rules of knowledge production). The fact that both thinkers discussed here, Bekker and Descartes, aimed to let God still be real in their versions of a disenchanted world is remarkable, but it could not prevent that later thinkers pursuing a further disenchantment expulsed God as well from the real world.

Kant and the boundaries of reason

Over the past decades, Kant scholars have shown renewed interest in the pre-critical Kant, and also especially in his little treatise of 1766 on one the most famous spiritual writers of his time, Emanuel Swedenborg. Interestingly, Kant's famous argumentations to limit valid knowledge within the bounds of pure reason, are tried out there to deal with the issue of whether to take spiritual revelations as Swedenborg described them in his books seriously. This shows that the banning of the spirits in modern culture is not the mere side effect of an independent development toward modern rational scientific approaches to knowledge, but that, on the contrary, the question of the acceptability of spirit ontologies was at the heart of Kant's most influential delimitation of valid knowledge. This makes the co-concurrence of the present returning interest to indigenous relations to nature, to indigenous knowledge and to spirit ontologies, with the growing impossibilities to control our world on a scientific and secular basis (mass dying of insects, warming of the earth, etc.) to be no coincidence. The moderns tried their approach, based on secularized knowledge and taking control, and now that it reaches its limits, other ways to relate to nature are being heard louder, as indigenous leaders and speakers increasingly find the media to propose their ways of being in the world as a more hopeful and sustainable alternative. One of the points I want to make in this book is to be wary of attempts by the moderns to appropriate elements of shamanistic relations to nature, while maintaining the frameworks of neo-colonial relations between modern and indigenous cultures. In order to avoid this, it is important to address the politics of epistemology, and to center the most important parting of the ways – the one that concerns spirit ontologies, which characterized the rise of modern culture. So let us address this issue in the relation between Kant and Swedenborg. I will approach Swedenborg as most philosophers do – by going into Kant's critique of his spirit philosophy first, and after that I will treat of his proper thought and his potential heritage despite being written out of Western philosophy.

Dreams of a Spirit-Seer (Kant 2002 [1766]) reads as an essay more than a systematical philosophical work. In it Kant satirically aims at two kinds of opponents: the dreamers of reason and the dreamers of the senses. The first-mentioned are the metaphysicians of his time (most notably Leibniz and

Wolff), who built conceptual systems of things which we cannot know anything of – the 'beyond' of physics. The dreamers of the senses are people like Swedenborg, who obviously have extrasensorial perceptions of a world beyond normal experience. That kind of experience is as such not the aim of Kant's critique, but the projections that Swedenborg makes on the basis of his personal experiences are. Swedenborg described the worlds of spirits, of heaven and of hell, in the most literal terms. He noted down how the spirits live, how they communicate with us, how their marriages work and more specific details that make the critical philosopher suspicious. Swedenborg also makes, in his work, very literal claims as to how he gets his information, mentioning spirits, angels, all kinds of non-human beings that appear to him, and with whom he converses as if with a living human being. Kant starts his essay in the most dismissive tone:

> The realm of shades is the paradise of the visionaries. Here they find an unbounded land, which they can cultivate at their pleasure. Hypochondrial vapors, wet-nurse tales, and cloister miracles do not leave them short of building materials.
>
> (Kant 2002 [1766]: p. 3)

Near the conclusive part, after having described and analyzed the principal ideas of those he characterizes as dreamers, we find Kant's privileged 'solution' to sensory as well as intellectual confusion – that we can, and should, draw boundaries, limits, that mark what we can meaningfully know and what we cannot. He concludes that

> ... metaphysics is a science of the limits of human reason; and since a small country always has long borders, in general more depends upon its knowing and guarding its possessions than blindly pursuing conquests ...
>
> (Kant 2002 [1766]: p. 57)

In his final chapter he draws similar conclusions, but now focuses on the moral merit of such land-surveyor work, on what he calls 'wisdom.' Kant admonishes his readers to be so wise to limit their pursuits of knowledge to include only those that indubitably found human morality, and to leave any unnecessary ideas behind. Among them he counts the idea that good men will be rewarded in the afterlife:

> ... it seems to be more in accord with human nature and the purity of morals to base the expectation of the future world upon the sentiment of a well-constituted soul than, conversely, to base its good conduct upon the hope of another world.
>
> (Kant 2002 [1766]: p. 63)

We see here a radicalization of what began with Bekker, who excluded demonic spirits from any valid moral reasoning. Like him, Kant places all trust in humanity itself, as we know it here and now, and claims that we have no need for any pseudo-knowledge of a beyond, a spiritual realm where human souls converse with other spiritual entities. All the same, Kant does not say anywhere that Swedenborg is a liar, his extraordinary experiences may be real, and not even nonsense. We just, thinks Kant, don't need them in any way to found the good human life upon, nor do they add anything useful to what we can know through empirical science.

Gregory Johnson in his introduction to the new translation of Kant's essay explains that it has always extracted ambiguous reactions from its readers. It has a strikingly different tone in its different parts, and the reader may wonder what Kant's aim actually is: was it really to make an end to the fame of the spirit seer and to attack in the same blow contemporary metaphysicians, as the accepted interpretation has been for a long time? Or was Kant's position toward Swedenborg more complex and ambiguous? He does write with respect and interest on Swedenborg on certain pages, and more so in other writings that have survived. Johnson tries to provide an interpretation of *Dreams of a Spirit-Seer* that could explain this ambiguity:

> I discovered that, in spite of the popularity of the received view, there have always been some skeptics who suspect that Kant's professions of scorn for Swedenborg are not the whole story and that along with, or behind, the scorn is a genuine respect for Swedenborg and perhaps even actual debts to his thought.
>
> (Kant 2002 [1766]: p. xiii)

The new English edition also contains passages from those other texts where Kant refers to Swedenborg, as well as reactions on the book from contemporaries, and the famous letter to Charlotte von Knobloch, that has a generally positive tone. This letter discusses the trustworthiness of the most famous examples when Swedenborg uncovered or predicted things in the empirical world on the basis of his contact with spirits, and describes how others confirmed them. He writes to von Knobloch how he wished he could have interviewed Swedenborg himself and how he awaits 'with longing' his next book, which he has already ordered in London (Kant 2002 [1766]: p. 71).

Kant scholar Friedemann Stengel has argued that there is more to this appreciation for the spirit seer. According to him Kant's exclusion of any spiritual, extra-sensory perception from the realm of valid knowledge goes hand in hand with his keeping many Swedenborgian 'metaphysical' ideas, where I mean metaphysics in the critical sense of Kant himself – as the discipline that decides on the boundaries of the world which we can meaningfully know.

... one cannot deny that one can find a whole collection of ... elements from Swedenborgs doctrine in his critical phase, to be precise in the sphere of the postulates of practical reason, as regulative principles or as objects of reasonable faith.

(Stengel 2008: p. 98, translation from the German A. R.)

More specifically Stengel holds Kant's recognition of two ontological realms (the noumenal and the empirical) by practical reason, a recognition he denies them from the standpoint of theoretical reason, to be a way to conserve, transform and embody an important piece of Swedenborgian metaphysical thought. What the meaning and importance of this piece is, I will discuss in the next section.

Swedenborg's heritage

Emanuel Swedenborg was a successful engineer in the service of the Swedish state mines, before he made his second career as a spiritual writer. Born as Swedberg, from a family of miners and pastors, he later changed his name to Swedenborg. From an early age, the spiritual life was important to him, although he actively pursued a career in technology. His father was a pastor, a theologian and later even a bishop, and a prolific writer himself. This father, Jesper, held a position toward religion that one should continuously strive to live according to the gospels. Like his son, later, the father was in close contact with spiritual beings:

Two original traits are discernable in Bishop Swedberg's pious personality: first, an emphasis on the practical side of the Christian religion and, second, a direct visionary experience of the influence of the heavenly world upon the mundane, an experience expressed in his belief in angels and devils.

(Benz 2002 [1948]: p. 4)

According to Jesper, the human relation to the divine takes place through the mediation of angels, serving spirits who help and watch over human beings. He also claimed that the souls of the deceased could still connect to living humans, and he practiced spiritual healing (named exorcism in Christian tradition – the expulsion of evil spirits). Like his father, Emanuel understood religion as a personal relation to the spiritual realm, a relation which came naturally to him already as a child:

From my 4th to 10th year, I was constantly in thought about God, salvation, and man's spiritual suffering. Several times I disclosed things that amazed my father and my mother, who thought that angels must be speaking through me.

(Lachman 2009: p. 9)

Besides his great interest in modern science and technology, he studied his dreams, practiced meditation through breathing techniques and wrote a personal diary to interpret what happened to him on the spiritual level. In his mid-life, however, he experienced a severe crisis, which made him doubt his view of life, and in which he was confronted with an acute awareness of evil powers. This period of inner turmoil finished in a breakthrough, a vision of Christ himself and a sense of a calling to understand religious tradition and the biblical revelations in a new way. He described his new state of consciousness as a semi-continuous opening of the spiritual realm to him:

> That same night were opened to me ... the world of spirits, heaven, and hell.... From that day, I gave up the study of all worldly science, and labored spiritual things, according as the Lord commanded me to write. Afterward the Lord opened, daily very often, my bodily eyes, so that, in the middle of the day I could see into the other world, and in a state of perfect wakefulness converse with angels and spirits.
>
> (Lachman 2009: p. 89)

From that time on Swedenborg spent his life trying to explain the radically new view of Christian revelation he gained in his condition of being spiritually awake. What sets his spiritual writings apart is their interest in metaphysics – the reality of the spiritual realm, and the primary way to interpret its meaning for us on the basis of one's own experience. His approach leaves the traditional theology that works from biblical quotes and Greek metaphysical concepts behind, and builds a new 'metaphysics' on the basis of personal experience. In this sense his work is truly modern, analogous to the seeking of certainty for our sense impressions in a personal reflection, as Descartes provided, Swedenborg provided a means to find certainty in religious matters in a method of interpreting one's spiritual experiences.

> The decipherment of its [of the Bible, A. R.] inner sense became the only path to understanding. But this 'inner sense' is not theology in the meaning of a system of biblical concepts. Rather it relates the teaching of the Bible to ideas endorsed by his visions and intuitions. He used the allegorical method to harmonize the content of his intuitive vision of the universe with the content of the Scripture.
>
> (Benz 2002 [1948]: p. 352)

The frame of his Bible interpretations is the idea that there is a correspondence between the divine, the spiritual and the natural realm. Every natural reality expresses a spiritual and a divine reality. This expression is itself a mark of the central force in everything: life, or love. Thus we experience God in the vital force that makes us experience all the living as individual persons, from our own uniquely individual free love.

> The life of influx is the life which proceeds from God, which is also called the spirit of God, of which it is also said that it illuminates and enlivens. But this life assumes a different form and changes according to the organization, which it receives through his love.
>
> (Swedenborg, cited by Benz 2002 [1948]: p. 367)

Although Swedenborg's insights and works are formed within a Christian framework, it would be interesting to ask whether they confer in one or more aspects a shamanistic way to understand and to be in the world. Of course his view of God, of spirits and of human life, is formed in a modern, not in an indigenous context. Most importantly he understood and communicated what he experienced as a new direction for modern religious life, a direction that would help people to be in touch with their spiritual life and personal love, instead of with ritualistic and moralistic commandments as the church handed them down. It is understandable that his work was of interest to Kant, as it is modern and philosophical in the sense that it provided an interpretation of reality that put human experience central – and while doing so attempted to explain issues concerning practical reason, such as the will, morality and our relation to finitude and death. In *Heaven and Hell*, Swedenborg states:

> Further, love is the essential reality of every individual life. It is therefore the source of the life of angels and the life of people here. Anyone who weighs the matter will discover that love is our vital core. We grow warm because of its presence and cold because of its absence, and when it is completely gone, we die. We do need to realize, though, that it is the quality of our love that determines the quality of this life.
>
> (Swedenborg 2010 [1758]: p. 11)

The reader who is educated in secularized philosophy has to read through a layer of strangeness to understand Swedenborg's spiritual writings. Swedenborg relates in a very literal manner how he has visited hell, and heaven, and has conversed with angels, deceased, people who have been taken up into these realms and with evil spirits as well. All the same however these works show a great consistency in their understanding of the human being and of life in its ontological and practical senses. The problem is that we are not used to the imagery he uses, in a modern, scientifically educated writer, while we may accept similar language when we read anthropological accounts of spirit ontologies in peoples who live outside or on the outskirts (like the Runa among whom Kohn did his research) of modern culture. It may be one of the reasons why Swedenborg has been marginalized and not been studied and interpreted very much. To my view the thorough academic study of his work, despite the efforts of many scholars who have worked on it, is still in his infancy – even when it is done as part of the 'history of ideas,' that is, in a more descriptive fashion. More so, however, when it is studied from a philosophical viewpoint. The problem is that we simply do not have an interpretive framework to understand texts that are modern and based on spirited experience likewise.

For Swedenborg as I understand him, the main ontological and ethical insights are, first, that there are spiritual worlds (heaven, hell and a separate spirit realm between our normal life and those others) next to the empirical world – it is not 'beyond' in the sense that one can only have access to it after death, or in extraordinary revelations. It is always already here, hidden only by an epistemological film that separates the material, technical relations we have to things and the spiritual, more essential way how things relate.[2] This more essential way is that everything is 'caused' from within, so to speak, from the divine source of love that breathes life into both worlds, the spiritual and the material. Heaven, and hell, are not thought of as distinct and independent from the everyday world – in fact what we love and think and do creates the spiritual body that makes us part of hell or of heaven after death. Also, as his own experiences confirm, we can be there even while alive in the mortal body, in a trance state. Heaven may be seen to consist in the consistent focus on truth in thinking and on good in volition, as hell consists of the opposite, as

> …no one is let into heaven until she or he is focused on what is true for the sake of good, and no one is cast into hell until she or he is focused on what is false for the sake of what is evil.
>
> (Swedenborg 2010 [1758]: p. 542)

Here we have arrived at the second most important insight in Swedenborg's works, that the spiritual world being the inner side of the material one, it works in the latter – our moral behavior being the outer phenomenal aspect of our being attracted by the good or by evil, of our love on the spiritual level.

> Even though we completely correspond physically to all of heaven, we are still not images of heaven in outward form, but only in inward form. Our deeper reaches are receptive of heaven, while our more outward ones are receptive of this world.
>
> (Swedenborg 2010 [1758]: p. 101)

Hell and heaven being understood as spiritual states of humans,[3] their balance is what we call human freedom. Whereas truth, power and the good is all from God, evil is human, and as we know both, being attracted by God as well as by evil, we have a free will, a conscious choice to realize the one or the other spiritual state. It is really a human spiritual, or psychological even, state, not a world beyond the clouds where we are cast into, or taken up into:

'The evil within us is hell within us, for it makes no difference whether you say "evil" or "hell"' (Swedenborg 2010 [1758]: p. 332).

The third important aspect of Swedenborg's thought is his idea of thinking, of the epistemic aspects of being human. Understanding everything, and mostly our own being, as flowing from God, is understanding the world from love, and in truth, while understanding things only from our own human being, or as purely worldly, is understanding falsely, and thus choosing evil over against God. What is relevant now for our purpose is to look through the Christian framework of Swedenborg's thought, and see how, within the context of modern culture, with its focus on the human being as the center of everything, in a moral, an epistemological as well as an ontological sense, it aimed to maintain an understanding of things from a deeper level of being spirited, seeing clearly that without such an understanding human personhood, truth and morality would become obsolete. As argued above, these aspects may have influenced Kant and made him keep to the free will as a regulative idea in practical life, even when theoretical reason cannot accept such a thing to exist in the world as science knows it. Free will in the Swedenborgian sense is however, in contrast to Kant, understood as the option to be led by love of God and truth, while Kant narrows it down to having understanding of the good as such, without relating it to its divine source. So, even though morality was (most precariously) saved in the confines of pure reason, it was detached from the spiritual, and spirit ontologies were excluded from the Western world of scientifically modeled knowledge production – making modern Western ethics a potential instrument for the exclusion of all but its own secularized approach. Thus it also could become an instrument for the exclusion of those human beings that it declared to be not fully reasonable or human.[4]

Notes

1 These are the words of the Iglulik shaman Aua, given to Danish ethnographer Rasmussen and cited by Harvey.
2 I only give a simplified rendering of Swedenborg's description of the different layers of reality here, that works for the purposes of discussing a spirit ontology in the context of modern culture.
3 Swedenborg thinks that animals have a spirit like humans, but theirs dies when their body dies, making heaven and hell uniquely human states.
4 Because of the crucial role of ethics in the Western othering of non-European, non-white peoples, it is not surprising to read in the recent book by Tommy J. Curry that 'Anti-ethics is necessary to demystify the present concept of MAN. It is an attempt to expose the assumed ethical orientation of reason as an essential anthropological quality of the human [...] as an illusion and a stratagem' (Curry 2017: p. 186).

Chapter 4

The return of (animal) spirits in the modern Western world

A wider experience

It seems that some core ideas of Emanuel Swedenborg did influence the critical works of Immanuel Kant, and more work must be done to further investigate his relationship to the Swedish spirit seer. Still, the effect of Kant's fame, the influence his work had on philosophy in Europe and later globally, resulted in the widespread defamation of all other than scientific knowledge and the expulsion of spiritual insights in the field of ethics. About one and a half centuries after Kant, however, we hear a voice in philosophy contesting this:

> There are resources in us that naturalism with its literal and legal virtues never recks of, possibilities that take our breath away, of another kind of happiness and power, based on giving up our own will and letting something higher work for us, and these seem to show a world wider than either physics or philistine ethics can imagine.
>
> (James 1996 [1909]: p. 305)

A most radical challenge was undertaken by William James to return to Kant and argue for an alternative approach of spiritual matters. This was done so thoroughly that we may speak of a return of spirits in philosophy and psychology. In his *The Varieties of Religious Experience* (2002 [1902]), considered to be the founding work of modern psychology of religion, James claimed that studying religious phenomena, especially the original religious experience that is primary to any organized religion, will make it possible to circumvent the rejection of spiritual realities in the sciences (including the humanities and social sciences). In American academic circles, obviously, religion was a serious subject to be studied and reflected upon, the reality of spirits was not. In a lecture, when addressing spirit phenomena, James noted:

> But those regions of inquiry are perhaps too spook-haunted to interest an academic audience, and the only evidence I feel it now decorous to

bring to the support of Fechner is drawn from ordinary religious experience.

(James 1996 [1909]: p. 299)

In short: religion is allowed, spirits are not. And thus he writes his great work on religion focusing on religious *experience* – and not on religious practices, ritual, theology, community and whatever is relevant in general to religious studies. The experience on which he focuses, is the place where the reality of spiritual phenomena finally may enter scholarly discourse once more.

In *The Varieties of Religious Experience* (2002 [1902]) we see how James can take spiritual/religious phenomena seriously, by putting his pragmatist ontology to work. Pragmatism claims that the right principle to decide whether something is real, is not in the observation that it *is*, but rather in the observation of how it *works*. Concerning religion this means that those beliefs that help to free a person from despair have religious reality, whereas those that only reproduce ideas without affecting the heart and mind of real people, do not. For James, ontology and epistemology are closely connected, to the point that they are almost impossible to discern – for in pragmatism it makes no sense to speak of realities that cannot be known, such as 'things in themselves,' or 'predicates of God,' nor of things that exist without humans relating to them.

In a third of the 20 lectures that make up *The Varieties of Religious Experience*, he also puts his pragmatist approach to use to critique Kant:

> Immanuel Kant held a curious doctrine about such objects of belief as God, the design of creation, the soul, its freedom, and the life hereafter. These things, he said, are properly not objects of knowledge at all. Our conceptions always require a sense-content to work with, and as the words 'soul,' 'God,' 'immortality,' cover no distinctive sense-content whatever, it follows that theoretically speaking they are words devoid of any significance. Yet, strangely enough they have a definite meaning for our practice. We can act as if there were a God; feel as if we were free; consider Nature as if she were full of special designs; lay plans as if we were to be immortal; and we find then that these words do make a genuine difference in our moral life.

> (James 2002 [1902]: p. 42)

Kant was right, according to James, in stating that human psychology is moved by any kind of 'metaphysical' idea it holds onto, such as the idea of free will, of God and of immortality. He was wrong to state that only concepts with empirical content can transfer knowledge of reality.

> ...this absolute determinability of our mind by abstractions is one of the cardinal facts in our human constitution. Polarizing and magnetizing

us as they do, we turn towards them and from them, we seek them, hold them, hate them, bless them, just as if they were so many concrete beings. *And beings they are,* beings as real in the realm which they inhabit as the changing things of sense are in the realm of space.

(James 2002 [1902]: p. 43, italics A. R.)

Here and in other parts of the book, James argues for the inclusion of beings in the realm of spirit into philosophical ontology, convincing his readers that Kant had it wrong. To do so, he needs the idea of a plurality of realms, or worlds, which he defends in a work he wrote after *The Varieties of Religious Experience* (2002 [1902]), *A Pluralistic Universe* (James 1996 [1909]). In neither of these two works does he ever mention Emanuel Swedenborg, but sometimes one gets the impression that would Swedenborg have had a more profound passion for philosophical argumentation, he would have responded to Kant as James did.

James-expert Eugene Taylor argues that there is more to the extraordinary fit of Jamesian ontology to Swedenborgian metaphysics. James, he claims, not only needed his pragmatist arguments to give a foundation to his psychological work on spirit phenomena as such (Blum 2006), but he was also pushed into finding a philosophical foundation for taking spiritual reality seriously by his father, who was (albeit for a short while) a Swedenborgian. Taylor describes William's father, Henry James Sr., as an '... errant, utopian socialist, ... Calvinist and later Swedenborgian philosopher of religion, who was an aspiring nineteenth-century literary figure in his own right' (in James 2002 [1902]: p. xvii). After James the elder suffered a spiritual crisis in 1844, a Swedenborgian minister helped him to recover. This led James Sr. to read most of Swedenborg's writings. In his introductory article to the centenary edition of *The Varieties of Religious Experience* (James 2002 [1902]), Taylor makes a case for the idea that William James not only lived in a circle of transcendentalist and Swedenborgian thinkers and writers, among whom the famous writer Emerson, but that more so, the influence of his father made him, unconsciously or consciously, develop his philosophy in such a manner that the spiritual realm could be included in what had to be considered reality. On top of that, Taylor lists several specifically Swedenborgian ideas that entered Jamesian pragmatism:

> Other Swedenborgian ideas taken up by the transcendentalists included the Doctrine of Use, which influenced James's later definition of pragmatism; the action of Divine Providence, which became James's later doctrine of tychism; the influx of divine power into the field of normal waking consciousness, which was James's later statement on mystical awakening; and the concept of rationality. This was not the mere rationality of the logicians, however; it was reason, based on intuitions and their visible effects in action.
>
> (In James 2002 [1902]: pp. xvi–xviii)

These historical observations on the influences on James's work may lead historians of thought to reconsider his interventions in philosophy, psychology and the study of religion, as is already being done to the works of Kant as well.

Where the most influential works of the Enlightenment thinker led to a general atmosphere in which it was deemed the right thing to discard with experiences of spirit phenomena (especially in the academic layers of Western culture), James worked very hard to give room once more to such wider experience. His pragmatist approach was in all he did the vehicle to bring unfiltered experience back into the spotlight. Where James focuses on religion in *The Varieties of Religious Experience* (2002 [1902]), he argues that the philosophy of religion, aiming to prove by reason alone the existence of God, can never lead to capture the original impressions people have of the divine, as those people do not care about such things as abstract theoretical proofs. What counts for them is the direct practical effects on the individual's mood, as well as on the actions one takes inspired by them: '... the uses of religion, its uses to the individual who has it, and the uses of the individual to the world, are the best arguments that truth is in it' (James 2002 [1902]: p. 321).

Because James was pioneering his new approach to spirit phenomena, he often uses the words of others to find the articulations he needs. To articulate how to think of religion in terms of pragmatic effect, of workings, he extensively cites the French theologian Sabatier:

> Religion is an intercourse, a conscious and voluntary relation, entered into by a soul in distress with the mysterious power upon which it feels itself to depend, and upon which its fate is contingent. This intercourse with God is realized by prayer. Prayer is religion in act; that is, prayer is real religion. [...] Religion is nothing if it be not the vital act by which the entire mind seeks to save itself by clinging to the principle from which it draws its life.
>
> (James 2002 [1902]: p. 325)

Another voice that comes to James's rescue when he aims to articulate the reality of the spiritual is his co-researcher into psychic phenomena Frederic Myers. In this kind of research, which they undertake for the newly founded Society for Psychical Research, phenomena are at stake such as mystical visions, sudden feelings of being guarded by something higher, etc. He cites Myers accordingly with approval:

> There exists around us a spiritual universe, and that universe is in actual relation with the material. From the spiritual universe comes the energy which maintains the material; the energy which makes the life of each individual spirit. Our spirits are supported by a perpetual indrawal of

this energy, ... and we must place our minds in any attitude which experience shows to be favorable to such indrawal. *Prayer* is the general name for that attitude of open and earnest expectancy.

(James 2002 [1902]: pp. 326–7)

James articulates these thoughts later in his own words as spiritual energy becoming active in prayer, and spiritual work being brought to effect by it (James 2002 [1902]: p. 334). Finally he concludes that a practice such as prayer, but also spontaneous spiritual experiences, as well as living a 'saintly' life, bring us in close contact with what he calls the 'transmarginal or sub-liminal region' (James 2002 [1902]: p. 338). This region holds what different theologies aim to define as God, or nirvana, but that in the words of the philosophical-psychological approach that James developed with Myers can better be described as a wider reality than the empirical one, a spirit reality, from which our individual self, and its consciousness, draws its life. What it is, this wider reality, is not so important for practical religion, as long as we experience it as higher than ourselves, better and more holy, so that it can inspire us in our actions in this life. The effects of our communi-cation with that reality take instance by working on the energy-centers of all the other subjects (James 2002 [1902]: p. 361).

It is important to note that his work on religion and on spirits did not lead James to favor a monistic or monotheistic ontotheology. In fact he came to the conclusion that a certain kind of polytheism was much more in accordance with its outcomes. To be inspired spiritually, namely, he found that we do not need any kind of Absolute Being.

> Anything larger will do, if only it be large enough to trust for the next step. It need not be infinite, it need not be solitary. It might conceivably even be only a larger and more godlike self, of which the present self would then be but the mutilated expression, and the universe might conceivably be a collection of such selves, of different degrees of inclu-siveness, with no absolute unity realized in it all.
>
> (James 2002 [1902]: p. 366)

This conclusion is of great importance for our investigation. James's decision creates a philosophical framework that allows one to treat the many varying spirit ontologies across religions and cultures as localities in this spiritual space without boundaries. James's pragmatism helps us to avoid the Scylla and Charybdis of *either* taking certain ontologies to be *irrational* or supersti-tious, *or* taking *all of them* to be just *cultural projections of human power rela-tions*. This means it offers a third way between either reductionism and relativism.

In a Jamesian frame we can accept different spirit worlds to be real, next to, and informing this material one simultaneously. Later he calls this frame the multiverse. The epistemological-ontological position that allows for the multiverse is pluralism. Pluralism entails that no absolute, all-encompassing 'block' can be thought without leaving the bulk of human experience out. Every collection of realities is still one among many others, and thus has an 'outside':

> Everything you can think of, however vast or inclusive, has on the pluralistic view a genuinely 'external' environment of some sort or amount. Things are 'with' one another in many ways, but nothing includes everything, or dominates over everything. The word 'and' trails along after every sentence. Something always escapes.
>
> (James 1996 [1909]: p. 321)

The conclusion of *A Pluralistic Universe* (James 1996 [1909]) allows us to think of Songhay cosmology, of Yanomami ontology, of Swedenborgian metaphysics or of the emanations of neo-Platonism – and in fact of ever so many other systems to relate things to each other – as a true explanation of things we experience, and still hold that none of these ontological frameworks will be able to capture *everything*.

This move of James has several revolutionary consequences for the main issues of this study. The first is that if nothing can capture everything, nothing can, as he says, dominate over everything. This means the end to what Mall and other intercultural philosophers call hegemony: the domination of the modern philosophical way to understand the world over all others, with the effect that colonization, use and abuse of any and all 'others' is allowed. The second equally important consequence is that, if all ontologies now are in the same boat, they will have to negotiate their relations to the world or to nature with each other. A third and surprising consequence is that, given pluralism, shamanistic or spirit ontologies in fact have been right all along with respect to epistemology – being reluctant to decide once and for all what is true – going along in reflection with how life develops itself in and among us, taking up new ideas (among them modern ones) as they seem fitting to life's realities, while not giving up their own, handed down and renewed, whole of practices and ideas. This seems to be the tenet of what James calls the 'each-form' of reality:

> In the each-form ... a thing may be connected by intermediary things, with a thing with which it has no immediate or essential connexion. It is thus at all times in many possible connexions which are not necessarily actualized at the moment. They depend on which actual path of intermediation but may functionally strike into: the word 'or' names a genuine reality.
>
> (James 1996 [1909]: p. 324)

So spirit, and spirits, may be real in dreams, or incarnated in bodies, such our own, or that of trees or jaguars. It/they may be experienced as the helpers of a loving God, as angels or as moments in which divine energy is flowing through something, a word, a breeze, a glance, a touch. They can also take on frightening and warning forms, as evil spirits, as enemies that challenge and more. That many forms of real may be captured (when we accept Jamesian pluralism) by a family of concepts in different languages, which can be negotiated between us without having to deny the legitimacy of any particular ontological system, as all ontologies have always been in continuous conversation with each other after all.

Deciphering anim(al)istic dreams

It is no coincidence that those thinkers who tried to make space for spirits in modern philosophy reached across the boundaries of disciplines. The methodological demarcation of modern disciplines reflects the divide inherent in the Kantian dualism of the empirical and the noumenal world. The new science of psychology was supposed to study the human being by empirical means, philosophy was to limit itself to reflecting on reason, logic and structures of cognition. The empirical human being is considered to be a distinct individual, closed in in his body. The mind, that favored subject of analytical philosophical anthropology, is understood mostly in a supra-individual sense, as 'universal reason,' or as expressing deep, grammatical/logical structures. Due to this division of labor, experiences of unboundedness – unboundedness to a single, bodily defined individuality, or unboundedness to the logical structures of reason, the kind of experiences that go together with spirit mediation – fall through the crack that divides. As we saw in the previous chapter, they have been othered and labeled as the result of irrationality, mental illness or a primitive, undeveloped mind. Those thinkers who did try to keep this wider experience on the academic table had to find ways to deny or overcome the divide between the intelligible and the empirical. William James did so by moving back and forth between disciplines. Carl-Gustav Jung, who cannot be absent in a study like this one, was trained in the new empirical psychology, but increasingly crossed its disciplinary boundaries to venture into religious studies, anthropology, theology and philosophy. Jung was explicitly interested in all kinds of spirit experiences – as well in clinical psychological practice, as in literature from the world's wisdom traditions. He studied the ancient Egyptian book of the dead as well as Zoroasthrianism, yoga and Western esoterism. While doing so, he still kept to his scholarly aim to provide a theory that could explain these experiences, as well as the practices to mediate the spiritual. His explanation, which he labeled depth psychology, is characterized by two main elements: on the one hand it has strongly theological overtones, where he designates the aim of his psychology to be individuation, the integration of what he called our dark side. On the other hand it has a pragmatic element, where he takes symbolic, psychic knowledge to be judged by its effectiveness with regard to the work of individuation.

In an essay he wrote late in life, in 1956 (Jung 2011), Jung explains how psychology and philosophy have to be transformed and work together to get our 'shadow side' in view once more. He deplores how philosophy has lost her original aim to search for wisdom, and how religion has lost touch with the needs of modernity:

> Our philosophy is no longer a way of life, as it was in antiquity; it has turned into an exclusively intellectual and academic exercise. Our denominational religions with their archaic rites and conceptions – justified enough in themselves – express a view of the world ... [that] has become strange and unintelligible to modern man.
>
> (Jung 2011: p. 40)

Modern science provides the framework for the life of modern man. He lives by knowledge that is based on tested hypotheses, that prove their value by their ability to technically improve the conditions of life. This is still the condition of life in modern societies. The media provide us daily with reports on what foods have been proven to improve health, what lifestyles may prevent depression and which economic predictions should guide what to study, where to put our money, etc. The scientific outlook, however, provides no answer to questions for meaning, and therefore always brings with it the danger of what Jung calls 'nihilistic despair.' The existential situation of modern people is described by him as a collective psychological illness:

> The rupture between faith and knowledge is a symptom of the split consciousness which is so characteristic of the mental disorder of our day.... If for 'person' we substitute 'modern society,' it is evident that the latter is suffering from a mental dissociation, i.e., a neurotic disturbance.
>
> (Jung 2011: p. 41)

To cure this collective illness, Jung directs modern, Western people to look into what they have suppressed, have forgotten, '... towards the foundations of his consciousness – that is, towards the unconscious, the only available source of religious experience' (Jung 2011: p. 49). Like James, he thinks that organized religion has lost its ways to bring effective spiritual mediation, with the result that many in modern culture suffer from existential disconnectedness or alienation from the deep layers of life. And like James, he searches for an answer to this problem in opening modern people's psyche up once more to original, direct, religious experience.

Whereas James aims to open up the wider human potential by means of a positive, life affirming, approach, pointing out how one can learn from non-Western traditions to tap in once more into the deep spiritual sources of life, Jung chooses a psychoanalytical focus on our dark side, including the 'sins' we brought to others. In this context he shows awareness of the evils of colonialism, which burden 'the white man.' These, and other great evils done by him, have to be faced to heal the illness of modernity.

> Quite apart from the barbarities and blood baths perpetrated by the Christian nations among themselves throughout European history, the European has also to answer for all the crimes he has committed against the coloured races during the process of colonization.
>
> (Jung 2011: p. 52)

It seems Jung is aware of the theoretical and actual struggles of his times very well, as this work was written only a few years after publication of Fanon's and Césaire's anti-colonial classics, and at the time the wave of institutional decolonizations had started to take place all over the world.[1] Still he addresses only the white man himself and identifies him without any issue by means of the concept 'mankind.' The Enlightenment idea of human progress, in which Europe has taken the lead, and the others lag behind, is still the frame of thought of Jung's work. The contents of non-European traditions are taken by him as primitive, earlier, phases of humanity. His psychology transforms and reinterprets these contents into 'knowledge' for modern man. The 'primitives,' according to Jung, distinguish themselves from modern man precisely in the fact that they have no reflection, no knowledge of what they do. Contemporary 'primitives' are thus taken by Jung to linger in a previous phase of human history.

> The fact is that in former times men *lived* their symbols rather than reflected upon them. I will illustrate this by an experience I once had with the primitives on Mount Elgon in East Africa.
>
> (Jung 2011: p. 120)

The said experience consists in the ritual of the East Africans to spit in their hands every day at sunrise and offer their spittle to this divine moment, which they call 'mungu.' Jung identifies their primitivity as their reluctance to reflect on this act, which they therefore (he concludes) do not understand.

> When I asked them what they meant by this act and why they did it, they were completely baffled. They could only say: 'We have always done it. It has always been done when the sun rises.' … What they were doing was obvious to me but not to them. They just do it, they never reflect on what they are doing, and are consequently unable to explain themselves.
>
> (Jung 2011: p. 120)

This reminds us of the attempts of the anthropologists of the ontological turn to explain practices observed in Amazonian cultures. The underlying idea mirrors Hegel's observation that what is unreflected, which is not taken up by the mind, is not yet truly human.[2] It is not yet historical, and it leaves the peoples who are still living by such practices behind in a realm without progress, defined by cyclical time. Here we see the politics of epistemology right in action. The othering of non-European peoples takes effect through the epistemological criteria modern Western culture has defined and declared to be universal. Only what is translated into theory, only what has entered the realm of abstraction, of reflection, is recognized as valid knowledge. The possibility that there may be *knowledge in practices*, in the body, in the way it dedicates itself to the divine act of the sun rising, is outlawed.[3] The political aspect of the modern delimitation of knowledge is overlooked, which allows the declaration of what is valid and what is not to do its othering and oppressing work, leaving white knowledge guarded as the only *real* knowledge. While Jung thus reproduces Western epistemological hegemony, he also denigrates religion, as much as this also implicitly claims that knowledge is present in the body and its movements, as it adheres to ritual, just like the spitting at sunrise of the Africans Jung met. Christian monks bowing at sunrise in prayer to their God, Muslims prostrating during prayer are only two (similar) examples of ritualized 'shamanic' elements in organized religions. These and other ritual observances relate the embodied spirit of this human person to 'God,' favoring those moments where the expression of the one connecting energy can be felt, which may be thought to re-establish the individual into the interconnected web of living beings through which this energy is passed along, as a gigantic electric relay.

Jung recognizes that the modern approach to knowledge has caused existential illness:

> Man feels himself isolated in the cosmos. He is no longer involved in nature and has lost his emotional participation in natural events, which hitherto had a symbolic meaning for him.... His immediate communication with nature is gone forever, and the emotional energy it generated has sunk into the unconscious.
>
> (Jung 2011: p. 135)

Still, although the immediate communication with nature is lost to (modern) man, it can be retrieved, according to Jung, through depth psychology. We can restore our lost relation to nature in a transformed, sublated, form, by analyzing and through analysis understanding the anim(al)istic spiritual contents of our dreams – as in dreams the unconscious still speaks to us, in all its complex symbolism.

> The symbol-producing function of our dreams is an attempt to bring our original mind back to consciousness, where it has never been before,

and where it has never undergone crucial self-reflection. We *have been* that mind, but we have never *known* it.

(Jung 2011: pp. 137–8)

I speak of ani(mal) spirits, because the importance of the body, its empathic potential, its ways to orient itself through movement and feeling – all the embodied ways of knowing, reflect that we, humans, are spiritual as animals, as living, moving creatures. Jung seems only to have a partial understanding of this, reproducing the favorizing of 'mind' and of theory over practice and wisdom.

Jung, in sum, offers only a partial healing to 'modern man.' First, he repeats the Hegelian hegemonic scheme which judges reflective knowledge to be the higher, more progressed and only truly human way to relate to nature. Second, his moral acknowledgment of the evils white man perpetrated to othered peoples is, in his approach, only meant to benefit the same white man – to help him overcome the alienation of himself, by becoming *conscious* of his dark side. Third, Jung doesn't acknowledge that the category of 'man' itself is oppressive to human beings in their plurality, men, women and those of other genders. Fourth, by ignoring the possibility that there is real knowledge in practices, he reproduces the othering of indigenous peoples, as well as of the plurality of ways to be human, to *know* and to relate to nature. If he would do so, he would also have to recognize that there may be healing for perpetrator and victim alike in practices of repair and retribution. And finally, he does not allow room for non-human 'persons' to be in intentional meaningful relations with us – thus leaving the anim(al)istic aspect of spirituality to be developed for those who came after him.

Deconstructing life and death

Among twentieth century Western philosophers, Jacques Derrida may have written most explicitly on the subject of spirits/ghosts. He even dedicated a book to it, titled *Specters of Marx* (Derrida 2006 [1993]). This book provides no anthropology, no metaphysics and no ontology, but instead works to open up the idea of *hauntology* – in the context of the global victory of neo-liberalism, after the 'end' of communism. The ghost of communism, however, as the ghost that haunted Marx, will always return, according to Derrida, as returning is what ghosts (revenants) do. Most unexpectedly he thus claims that the spiritual, conjured away by the moderns, now returns in the most unexpected realm – that of political economy. In the subtitle of this work, he points prophetically to three things that keep late modern people in their grip: *The state of the debt, the work of mourning and the new international*. Debt – mortgages and loans and all the financial products that derive from them, have shown after Derrida to be able to destroy our societies. The nation states are fairly powerless over against the world of finance.

At a deeper level, there is the debt that modern riches have to the poor, who work for almost nothing, or for nothing, as slaves. The late moderns are also haunted by the new international, which is not an organized workers movement like the old one, but consists of networks of hackers, leakers and the undocumented migrants that with their numeric pressure test the idea of borders, citizenship and civic rights to the core. And finally, in an unexpected reminder of the work of African shaman Malidoma Somé (Chapter 1), Derrida stresses that the dangers of trusting the modern state to provide for human life and flourishing will go unaddressed as long as we forget the importance of mourning – relating to the dead to which we are related. Mourning is a work, according to Derrida, a practice of the body, when we actively mention the name of the deceased, inscribe it into a text, speak to him as if he is here, attract him to that effect to teach us about life and how to live.

The specter at the center of the work of mourning advocated by Derrida is of course Marx, the author (together with Friedrich Engels) who so famously introduced spectrality into European discourse on political economy in the opening sentence of their *Communist Manifesto*: 'A spectre is haunting Europe – the spectre of Communism' (Marx and Engels 2015 [1888]: p. 1). Let us cite from the introduction of the *Communist Manifesto* somewhat more extensively, to then discuss what Derrida does with Marx and consequently succeeds to write the most explicit philosophy of spirits since Kant's rejection of Swedenborg.

> All the Powers of Europe have entered into a holy alliance to exorcize this spectre: Pope and Czar, Metternich and Guizot, French Radicals and German police spies.... Two things result from this fact: I Communism is already acknowledged by all European Powers to be itself a Power. II It is high time that Communists should openly, in the face of the whole world, publish their views, their aims, their tendencies, and meet this nursery tale of the Specter of Communism with a Manifesto of the party itself.
>
> (Marx and Engels 2015 [1888]: p. 1)

What springs to the eye in this quote is Marx and Engels's assessment that what is acknowledged by the existing institutions of power as a danger, even if it does not yet have a clear face or form, nor a manifesto that states its mission and objectives, is already a power. Marx and Engels, known for their positivist approach, want their readers to leave the state of a declared spectrality and show themselves, articulate their aims, take on a clear form, so to speak and come out of the mists in the sunlight, to work effectively for the realization of communism. Positivism in the Marxist sense is out to destroy the spectral, by transforming it into a tangible, active force that can effectively work in the world of humans. In this sense, Marxism is modernist – it trusts that good will come, for human societies, from political actions that are founded on a clear, unified, rational will. Marxism as a science will provide the knowledge to undergird the necessity and the hoped-for effects of the revolution.

Derrida now takes this figurehead of positive materialism, Marx, and deconstructs it – drawing attention to the ghost of Marx as the father of communism, who was exorcized by neoliberalism after the fall of the Berlin wall in 1989, as well as to the ghosts that plagued Marx himself, that drew his attention – the specters of those who were uprooted and dispossessed by the industrial revolution, the poor and the oppressed that always return. The effect of his deconstruction is to make the reader aware that, while we can conjure Marx the positivist away, the 'spiritual' Marx will still come to haunt us. This 'spiritual' Marx (my adjective) is the one who calls for revolution in the name of justice, something that stands in tension with the ideas articulated in his work on capital, that economic laws govern the development of our societies. When laws are in place, revolution would not be necessary. Revolution lives from something that escapes whatever economic science can positively describe – a haunting discomfort about how existing policies address the conditions for human life and flourishing, a discomfort that lives from an awareness of justice, that the ghosts of the dispossessed, the maimed and the murdered instill on us. This discomfort is so 'spooky,' so strong, that people will be ready to do the most dangerous thing: overthrow the structures that keep chaos at bay. The 'new international' that Derrida calls up is characterized by that very same spookiness, as it is embodied by those who in the reality of an administrated world do not fully exist.

The relationship between spectrality and justice is further elucidated by Derrida in the story of Hamlet, that takes a central place in the book. Hamlet is addressed by the ghost of the dead king, his father. This ghost tells him that he was killed by his brother, the present king, now married to his widow. The knowledge thus brought to Hamlet, places him in an unsolvable dilemma – to call out the murderer, at the expense of being declared crazy (having listened to a dead person), or to be silent, and thus carry his 'craziness' with him for the rest of his life, having a knowledge that is unspeakable. This dilemma is, following Derrida, '... the question of the specter [...], whether it be Hamlet's or Marx's' (Derrida 2006: p. 9). To pose this question is to do the work of mourning. Of mourning he says:

> It consists always in attempting to ontologize remains, to make them present, in the first place by identifying the bodily remains and by localizing the dead (all ontologization, all semanticization – philosophical, hermeneutical, or psychoanalytical – finds itself caught up in this work of mourning ...).
>
> (Derrida 2006: p. 9)

Later in the text he also speaks of 'messianic remains' – making the connection between eschatology and revolution even more pregnant. Derrida's distinction between teleology and eschatology is meant to highlight where justice comes into the world and challenges modern dreams of making a perfect society. It rejects the idea of political communism that an ideal condition for humanity (the end of all ownership) can be defined as the telos of history, a final end that would draw human societies toward itself by inescapable revolutions. According to Derrida, we can know the specter that speaks to us and reminds us of justice – but this doesn't provide us with a clear goal that would be lying beyond the horizon and to which we should strive. It is all much more ambiguous, and the concept of eschatology testifies to that. Eschatology is confining oneself to the hope that justice will reign 'in the end,' without knowing when and where the end will be, whether it will be at the end of history or outside of time, perhaps here already, in another dimension. We can neither know what it will consist in exactly. Thus, Derrida unhinges the clear-defined ethical inspiration of human agency as it is understood in modern culture, especially where it concerns political agency. We are responsible, we are called upon by the oppressed, but all one tries to do, in order to restore justice, will be ambiguous – as justice is always beyond anything one can do, always calling for more. More recognition, reparation, attention; more humility, hospitality, friendship.

In the expression 'messianic remains' Derrida plays upon the Jewish confidence that waiting for the messiah, who will restore justice, is the thing that renders human life meaningful. This waiting knows no time and place, and this is to be taken in the most radical way: we are, living here and now, indefinitely responsible for our actions. In the Derridian stress on our indefinite responsibility, one can detect the influence of Levinas's writings on justice, which is 'beyond,' transcendent in an absolute manner. Being ethical means listening to the radically transcendent justice that bends human agency, that makes the straight orientation of means and goals curved toward a goodness that escapes our normal reality. Whereas Levinas speaks about 'ethics before ethics' in the abstract language of Husserlian phenomenology, Derrida relates our knowledge (of us, moderns) of responsibility to bodies – spectral bodies that remain to haunt us, thus bringing modern Western culture and thought down from the idealistic sphere of reflection to the concrete world of action. Marx himself was still a child of the Enlightenment that had banned the ghosts:

> ... Marx does not like ghosts any more than his adversaries do. He does not want to believe in them.... He believes he can oppose them, like life to death, like vain appearances of the simulacrum to real presence.
>
> (Derrida 2006: p. 57)

However, the ghosts/spirits return *qualitate qua*, they are revenants, so no exorcism will ever be final. Exorcism can never be annihilation, but only fending off, removing to a distance, to get some temporary calm, so that we can do those things that we know are morally problematic. Derrida does not link his analysis of the specters of modernity to spirit ontologies of indigenous peoples. More so, the spiritual does not fit the categories of ontology, according to him, but comes to us as hauntology, because we have defined ontology in such a manner (on the basis of the principle of the excluded third) that apparitions of *non-beings* that still *are* in so much as they appear (take on bodily form), cannot be taken to be real. Hauntology, therefore, is the shadow side of modern Western ontology.

Allowing the idea of hauntology, now, deconstructs the opposition of life and death. The living are those who want the dead to keep out of their world. Dead are not only the deceased, but all those spirits that are not alive in the embodied manner that we call life, experience to be life – so perhaps even my own spirit in so far as it escapes the confines of my body. The spirits that haunt continuously put the boundaries of life at risk, they bring the risk into our lives that the dead are not completely dead and that therefore, perhaps, we are not completely alive. Therefore 'effective exorcism pretends to declare the death only in order to put to death. As a coroner might do, it certifies the death but here it is in order to inflict it.' The certification '...speaks in the name of life, it claims to know what that is. Who knows better than someone who is alive? It seems to say with a straight face' (Derrida 2006: p. 59).

What takes place in such exorcism, Derrida moves on, 'is a way of not wanting to know what everyone alive knows without learning and without knowing, namely, that the dead can often be more powerful than the living' (Derrida 2006: p. 60).

What is interesting for us, now, is that Derrida, avoiding the idea of culture, of anthropology or ethnography, even of ontology, circumscribes the experience of the spectral in such a manner that we are reminded of 'indigenous' ways to be in the world. The interconnectedness of morality, of danger and of conjuring rituals; the necessity to ritually reinstitute the orderly distinction of life and death, and of the society in which humans can live in peace and the wilderness in which the spirits haunt;[4] and finally the awareness of a continuous link beyond life and death that moves history. Ancestors tell shamans the directions which give the living the opportunity to live healthily and meaningfully. And in Derrida's work the unmourned dead of modernity indirectly tell us our direction. Indirectly in so far as moderns don't ask shamans to mediate. Therefore the spirits come disorderly, inspiring chaotic popular movements for change. Even more chaotic as in times of the old international; now that communism, and Marxism are 'dead.'

To make the return of spirits in philosophy possible, Derrida does not take the positive route of James, who aimed to describe spirituality as a transculturally accessible human characteristic. Derrida takes the negative route, to investigate where there is room in the margins of the hegemonic discourse of modernism. Where modernism has left cracks in its edifice through which spirits can move. In so doing he has to reintegrate actual individual beings in philosophy, not 'the body' of Merleau-Ponty, but actual bodies of mourned friends. In the dedication of *Specters of Marx*, one is mentioned by name – Chris Hani, who fought apartheid in South Africa and was murdered shortly before the book was published (Derrida 2006 [1993]). Of Hani Derrida writes:

> But one should never speak of the assassination of a man as a figure, not even an exemplary figure in the logic of an emblem, a rhetoric of the flag or of martyrdom. A man's life, as unique as his death, will always be more than a paradigm and something other than a symbol.
>
> (Derrida 2006: p. xiv)

He muses about the meaning of the murder of Hani, which was so difficult to understand. The murderer, a Polish emigrant, claimed to have killed him because he was a communist. Hani had just given up his role in the African National Congress (ANC) to join a minority communist party with hard to pin down intentions. He was therefore not killed as a figurehead of a powerful movement, but rather, the dedication suggests, of the democratic force showing in the 'luxury' people thought they could take in joining splinter parties, in democratizing in the sense of proliferating. As always, Derrida's text is here rich in suggestion: the suggestion that the recent abolishment of apartheid in 1990 was not the final end of a history of injustice, that so many hoped it to be, and Mandela was not the messiah, who finally eradicated the last remaining explicitly racist political system. The problem of humanity would remain. And the specters of those dehumanized by racism would come to haunt. Most importantly, Derrida seems to say that the new neoliberal order does not in any way mean the end of the colonial order, the modernist hegemony that others and dehumanizes those that it now declares to be 'underdeveloped' (the new word for 'primitive'). The gospel proclaimed by the neoliberals was that the free market would solve all problems and bring happiness and health to everybody, was the belief, and that the 'others' would gradually gain access into the happy white world by bringing them up to the levels of theory-based education, consumerist spending and multiparty democracy the West boasts. The white world itself was not challenged, in the ending of the Cold War, with the consequence that it should expect to be plagued by the eternal revenants of its victims.

The spirits return

Jung, James, Derrida – three thinkers that reached out from within the framework of modernism to try to capture the return of the spirits. If we follow Derrida, this return is necessary in order to 'learn how to live, finally.' Without an awareness of the workings of the spirit realm in our lives, we cannot get to what is truly human: to live in expectation of justice, and to expect it by working to restore it, that justice that never has been – that always has been beyond reality. Leaving the modern conception of a single, discriminatory, ontology – bounded in by reason and controlled experience – in place, Derrida suggests that this ontology is porous, and that ghostly appearances move through its cracks. These specters abide at the boundaries of life and death, of absence and presence – when someone dies, their body becomes ghostly, a 'zombie,' a body that tricks us, for a moment, in making us believe it can rise and walk and speak again. The trick becomes more serious, when even after burial or cremation, when the body has been 'put to death' once more, has been certified as dead, as dead within the society of the living, the dead person appears again, visually or auditory, or even tangibly as when a ghost rattles someone's bed or moves objects.

The issue that has been raised in the late modern condition of the 1990s by Derrida, is also the issue of animality – of life, of what it is to be alive, to be able to breath, to move, to feel, to fear. His analysis of the political-economic condition of 'our' time leads to the suspicion that a person with responsibility, an ethical person, a true human according to the principles of modern anthropology, cannot be in that state without the animality that makes it fear, feel and move – the animality that was expulsed from ethics and from true humanity by a certain Kantian approach. The animal body that notices movements, changes in atmosphere, is where we get into contact with the spirit world. For those who have been educated in the framework of mind-body, spirit-matter dualisms, it is counter-intuitive to realize that spirits are not immaterial, and that therefore they do not converse with other spirits unmediated. They communicate through our animality. That explains also that non-human-animals often play an important role when spirits appear: the cat's hair stands straight when the ghost passes by, the dog howls when he comes, rodents are nervous. In fact, it is the animal spirits, as they were called in the suppressed streams of Western thought, that pick up everything the senses don't. No return of spirits, therefore, without a new consideration of the animal nature of humans, therefore, pointing forward to the next chapter.

Jung held that the spirits that hold 'primitive' people in continuous fearful constraint to intervene in nature, have never left modern man completely. They show themselves, however, in modern culture as neurotic desires:

> The gods and demons have not disappeared at all, they have merely got new names. They keep him on the run with restlessness, vague apprehensions, psychological complications, an invincible need for pills,

alcohol, tobacco, dietary and other hygienic systems – and above all, with an impressive array of neuroses.

(Jung 2011: p. 122)

Thus our bodies bear the traces of the rejection of the spirits, they are still there in a negative manner – as illness, as sadness. Jung doesn't refer only to conditions that we would call psychiatric illnesses, but to a general condition of people who live in the modern world.

James relates spiritual reality to health as well, and thus to the human being as a living, embodied personality, immersed in networks of relations to the other beings in nature. To his view the new religious movement of 'mind-cure,' popular in his days, proves the 'primitive' ways to relate to nature to be right, next to whatever methodologically founded and universalistic claims the sciences may make to their impossibility and irrationality:

> But here we have mind-cure.... Live as if I were true, she says, and every day will practically prove you right. That the controlling energies of nature are personal, that your own personal thoughts are forces, that the powers of the universe will directly respond to your individual appeals and needs, are propositions which your whole bodily and mental experience will verify. And that experience does largely verify these primeval religious ideas is proved by the fact that mind-cure spreads as it does, not by proclamation and assertion simply, but by palpable experiential results.
>
> (James 2002 [1902]: p. 88)

It is, therefore, as animal, as a feeling, moving being, a being that continuously seeks balance to remain alive, physically as well as emotionally, that the human being relates to the realm of spirit. It is therefore understandable that those who still live close to non-human-animals see this kind of spiritual balancing act which is life as a common adventure in which all life is wrapped – and of which we can share much with those living beings close to us.

In shamanistic practices animals play an important role. They appear to guide human beings, to tell them things and also may offer themselves as food. This last 'spiritual' event is described by Felicitas Goodman, when she recounts how she and her students experiment with trance journeys while taking postures depicted in ancient statuettes. Her research describes how commonalities in the trance experiences convince them of the effectivity of the specific posture they try out. After one of such sessions in which they take the posture they identify as the 'hunter posture' she describes an unexpected meeting, upon going outside with her German shepherd dog, a dog that was normally patient and well-behaved:

This morning, however, she was restless. She went over the edge of the hill, stopped by a juniper bush [...] and started to growl, her hackles raised. Curious, we followed her, and there under the bush, nearly hidden by the branches, sat a porcupine. It was completely motionless, and as silly as this may sound, I swear it was grinning at us.

(Goodman 1990: p. 64)

The meeting is curious, as porcupines normally feed at night and don't show themselves to humans during daytime. It convinces Goodman that the porcupine most probably had reacted to the call of the hunter posture, '... and since it was willing to sacrifice itself, it was probably bewildered and pleasantly surprised that it did not end up on our spit' (Goodman 1990: p. 65). What is so attractive and special of Goodman's descriptions of her research and what she and the others experience during it, is that she combines a systematic investigative approach with an orientation that reflects her natural tendency to take shamanistic phenomena literally, as they present themselves. At this point we are reminded of her words in Chapter 1, where I rendered her description of the sadness that came over her when she realized that, growing up, the 'world of magic' had to disappear. She felt later in life that she was called to restore this part, not by way of a return into childhood, or by getting into the world of literature and poetry (that does allow the magic to a certain extent, as we saw in Hume, Chapter 2) or to turn to religion, but by investigating the natural potential she felt, to do 'shamanic' things. To organize trance sessions through drumming, explore the effect of taking certain body postures depicted in ancient statuettes and investigate what kind of knowledge could be found in the 'trance travel.' Apart from academic investigations into spirit realities, as the ones described in this chapter, over time there has always been an undercurrent of popular spiritual practices and consultation of mediums. Deborah Blum describes those in the times just before and during the adult life of William James. In earlier times, they are more likely to be found in groups and sects that functioned at the margins of organized religion, a phenomenon that also exists today. Today we see the spreading of neo-shamanist groups that grew out of the hippie-movement and alternative religious approaches. Often too neo-shamans and traditional shamans meet, in this age of global travel, at shamanic meetings and festivals. The epistemological and moral consequences of such meeting are still to show themselves.

Next to the ongoing esoteric movements, there are two phenomena that mark what I call a return of the spirits in the modern Western world. The first one is the hesitant beginnings – as shown in the authors discussed in this chapter – of serious academic interest in spirit phenomena. This shows that the spirits return to the realm of public awareness, that they move from more or less secret, private interests to something that has to be taken seriously by the learned world. The other phenomenon reflects that while in the age of empire, European (and white northern American) colonialists aimed to spread modern culture across the world, denouncing shamanistic cultures wherever they came across them as primitive, backward, semi-human ways to live, we now see a move of migrants from the former colonies to the countries of their former colonizers – and among those migrants there are many who bring with them some spirit practices that have survived in their hybrid cultures – making spirit phenomena, slowly but surely a new element in the North.

Notes

1 *Black Skin, White Masks* was published in French in 1952 (Fanon 2008), *Discourse on Colonialism*, also in French in 1955.
2 For a critical discussion of the relevant aspects of Hegel's thought, see Kimmerle 2014.
3 Of course it may also be the case the East Africans did not feel to open up their knowledge to a fleeting visitor like Jung. My critique here focuses on the problems inherent in Western delimitation of valid knowledge as such.
4 Instituting these distinctions is the primary job of the shaman, according to both Somé and Stoller.

Chapter 5

Deconstructing or decolonizing the human–animal divide

Deconstruction versus decolonization

The question whether 'man' (as the human species used to be called before long) has a special position in nature – as a quasi-outsider, able to control nature thanks to his unique ability to objectify the world around him – is central to the struggle over the environment. The view of man as the 'master and owner of nature' has become dominant in modern Western culture ever since the time of Descartes writing those words. It was articulated in a pregnant manner by Immanuel Kant in his *Anthropology from a Pragmatice Point of View*:

> The fact that the human being can have the 'I' in his representations raises him infinitely above all the other living beings on earth. Because of this he is a *person*, ... i.e., through rank and dignity an entirely different being from *things*, such as irrational animals, with which one can do as one likes.
>
> (Kant 2006 [1798]: p. 15)

Its effects reached a whole other level in the nineteenth and twentieth century, when industrialization made it possible to sustain ever larger habitations of humans, in a way that made it ever easier for them to forget their entanglement in nature as a whole. In our present age, we seem to have reached the turning point in the reign of this idea, now that climate change, fossil fuels, deforestation, desertification and the loss of animal species are everyday concerns. The limits of humanity's manipulation of the natural environment to serve its specific needs are coming in sight.

Simultaneously, and not coincidentally, the subject of animality is gaining attention in philosophical reflection in new ways – as in the rise of fields such as animal ethics and animal philosophy. Philosophers show more awareness of the likeness of humans and non-human-animals, shown in the fact that all of them can suffer, and thus attract empathy. Animal researchers show increased interest in investigating the intelligence of animals, their use of language, social behavior and even moral behavior. This is however still far removed from indigenous ways to relate to animals, which not only take them, in different ways, to be related to us ontologically (Bamana 2014; Kohn 2013), but that take animals, animal spirits or spirits who take the form of animals, to be bearers or bringers of shamanic knowledge. Also, most 'new' shamans, those who turn to shamanism within the context of modern life, such as Goodman, relate of the wisdom and healing brought through animal spirits – such as the 'Bear-spirit' which guides patients through a healing trance (Goodman 1990: pp. 100–15). In the shamanistic worldview there is no clear break between human life and all other life, resulting in the conviction that care for the environment is essential for every living creature, as it sustains all in their interconnectedness. This thus makes such care intrinsic to being, and not an afterthought to technology, as in modern cultures that see technology first and foremost as a means to stretch the ways of nature to solely suit human needs. Indigenous cultures take the animality of human and non-human-animals alike not as something to be suppressed to make room for rationality, but rather as a form of being spirited as such. Spirit is not reduced solely to human spirit (understood as reason), while non-human-animals are respected for their own kinds of knowledge from which humans can learn.

The modern Western tendency to experience the split between man and nature most strongly where we are closest to it – in our animality – led to a deep ambiguity in the treatment of animals in modern cultures. Some are privileged as house-animals who are treated almost as humans, and increasingly so. In rich Western countries animals get medical treatment that is often just as highly developed as that for humans. They may get psychological help, may own their own passport for travel on holidays with their owners, and get buried or cremated with dignity. Others, domestic animals that are used for their meat, skin and bones, are more and more separated from the house-animals, closed in in factory-like environments, where they are 'grown' and 'harvested' on a mass scale. This is far removed from the cows in old-fashioned farms that still had a name, and were to some extent treated as having the right to enjoy life. The third category, 'wild' animals, meaning those that are deemed unfit to be used for our material needs, or to be cuddly company, are treated with the highest level of ambiguity. On the one hand they may be controlled and managed, i.e., fed or shot as humans think their numbers are declining or growing too fast. On the other they are subject to romanticized projections of the true, untamed life on this planet, of which modern humans can only have melancholic dreams.

On the other hand, a similar ambiguity characterizes the ways in which modern Western humans treat the 'others' of their own culture. Even though explicit racist articulations are not accepted in academic discourse anymore, nor in the public expressions of representatives of organized religions, policies of modern nations may treat those who do not fit into the image of white-modern culture with an implicit or explicit racism that has not changed essentially since colonial times. Direct physical brutalities as in the times of slavery are not accepted anymore, but they are neither in the justice system for the privileged race. As Foucault has shown in his work *Discipline and Punish* (Foucault 1991 [1975]), after the eighteenth century, power works through disciplinary systems instead of through brute physical violence, but still can be just as oppressive. It works through institutions such as social housing, imprisonment, labor agreements and taxation. It exerts its influence from a center to peripheries, privileging or oppressing by means of different overlapping groupings according to class, gender, race, religion and culture. Those at the bottom are in all cases the 'indigenous' peoples, who are deemed to be no 'real' men, because they don't order their societies as moderns do. Enlightenment thinkers labeled them half-men, dehumanized and racialized them as 'almost animals.'

But even today modern disrespect and dehumanization shows in the careless destruction of their habitats, their culture, their family structures and so on. Destruction that is most often protected and backed up by legal rulings, as can be seen in the most prominent struggles of indigenous peoples, such as that against the Bela Monte Dam in Brazil, the Dakota Pipeline in the USA[1] and countless others that try to counter deforestation, mining and industrialized agriculture, and often don't even get into the news. Such dehumanization has been tracked and critiqued by postcolonial thinkers in Enlightenment philosophers such as Kant and Hegel, and will be discussed in this chapter. Simultaneously the ambiguities in the modern conception of animality have been tracked down in deconstructive work too. A deconstructive approach, as proposed by Derrida, as well as a decolonizing one, present in the work of Fanon and Eze, both problematize the concurrency of racist and speciesist othering of the non-human-animal and the animalistic nature of the human being. Although neither approach focuses on alternative ways to understand humanity and animality (as they can be found in indigenous worldviews), both are needed as critique – to blur the dividing lines that functioned to keep the possibility of spiritual encounters between human and animal out of 'real' philosophy. A (much needed) side effect of their critique will be the decentering of the human being and thus the decolonization of Western philosophy as it understands itself since Kant as anthropology. Ever since Thomas Hobbes, who claimed that 'natural men' were 'like wolves' to one another, Western philosophy has valued the animal aspects of our humanity negatively, and has seen the growth of reason and civilization as the way to overcome them. The relations between different peoples were understood in a similar vein: the Western, white, man was considered to be endowed with reason, while the 'others' were seen as almost 'animals' – almost, as Kant and Hegel saw beneficial effects in colonizing them, and leading them out of their supposed natural state into (Western) history.

Deconstructing animality

The deconstructive approach to human–animal relations was undertaken most explicitly by philosopher Jacques Derrida (1930–2004) in a conference address in 1997, of which the first part was published as 'The Animal That Therefore I Am (More to Follow)' (Derrida 2002). In a subversion of the usual attempts to demarcate humans and animals as belonging to clearly separate ontological domains, he starts out by investigating the animal aspect of being human – which results in a proposal to 'undefine' the concept of animality, as well as, consequentially, humanity. Such an approach is characteristic for Derrida, who in all his written work shows how modern philosophical understandings of reality have sought pure, discrete, separations, suppressing the contaminated or blurred boundaries we experience in actual life. He brings this suppression effect to light, not by building a critical argument against classical or mainstream thought, but by taking contaminated, deconstructive pathways while reading classical texts, thus changing the practice of thought.

From early modernity on, in the works of thinkers such as Hobbes and Descartes, philosophy has understood the human being as a thinking identity – as, essentially, disembodied intellectual spirit – meanwhile putting the animal on the other side of an imagined fence, as something with which we are not in contact, and by which we should not be contaminated. In the worldview of the moderns the state of nature, which is indifferent to the happiness of man, has to be overcome by instituting a lawful state and government. The passions of the body have to be brought under the reign of morality, and our physical suffering has to be outlawed by technology and medical science. Since we can do all of this, modern thinkers seem to hold, we are in fact always already beyond the natural condition and beyond animality. With our reasoning faculty we have been able to conjure it away, so to speak. The non-human-animal, as well as the animal-human body and inclinations, are both redefined in early modernity as essentially mechanically operating matter in motion. On top of that the non-human-animal is conceived as unable to suffer, and therefore not taken into account by moral philosophy.

Derrida now disregards this intellectual fence and sets out to discover our 'being with' animals – starting with giving back to the animal the gaze which modern man ascribes only to himself:

> The animal is there before me, there close to me, there in front of me – I who am (following) after it. And also, therefore, since it is before me, it is behind me. It surrounds me. And from the vantage of this being-there-before-me it can allow itself to be looked at, no doubt, but also – something that philosophy perhaps forgets, perhaps being this calculated forgetting itself – it can look at me. It has its point of view regarding me.
>
> (Derrida 2002: p. 380)

In a short but intriguing study, Patrick Llored has made an effort to reinterpret the work of Derrida as, in effect, an *ongoing* attempt to think animality. According to Llored, the question of animality is not purely philosophical, but rather an existential matter to the philosopher of difference. Derrida has remarked, in the above-mentioned article, that his entire work should be understood as originating from his sense of being 'the secret chosen one of those who are called animals.' Llored shows that early experiences of living in Algeria, especially in its 'Vichy' period, which led to the expulsion of the Jew Jacques from school, formed the source of Derrida's discovery of the link between racist and speciesist repression. And of its counterpart: the vulnerability of all living beings to violence. To cite Llored:

> Our hypothesis is that we should inscribe the Derridian itinerary in the movement of thought that has, in the preceding century, identified antisemitic violence and speciesist violence against animals, and which has united a large number of writers and important philosophers, having marked our culture of which it has revealed the intimate and deadly relations it holds over against animality.
>
> (Llored 2012: pp. 10–11, translation A. R.)

The young Derrida felt repulsion at being forced into a school reserved for Jews only, because in his eyes it reproduced the repressive violence toward Jews as a group. Here Derrida's critical thoughts about culture and nation as identifiers originated. Over the course of the years, in works such as *Acts of Religion*, he expressed his view that belief in political sovereignty, national unity and the identity of culture, as markers of the idea that 'man' could own himself, is a metaphysical belief at the root of communities (cf. Llored 2012: p. 15). Thus, exposing the intimate linkage between metaphysics and politics, as well as between the idea of an autonomous knowing subject and the suppression of non-human-animals, became important focus-points in his work. Here the different themes in his work connect as a full-fledged thinking *against* the Enlightenment idea that all philosophical questions in the end can be reduced to the question concerning 'man' as the uniquely free and rational being – the idea that

> ... to know the human being according to his species as an earthly being endowed with reason especially deserves to be called *knowledge of the world*, even though he constitutes only one part of the creatures on earth.
>
> (Kant 2006 [1798]: p. 3)

A quote like this shows that modern anthropology is actually a vindication of treating all and any potential non-humans as objects to satisfy human needs. Modern anthropology is therefore political and metaphysical – declaring the deep structure of reality to lie in the reason man is endowed with, which in its turn is allowed to suppress any other claim to subjectivity, personhood, of any non-human-animal.

In 'The Animal That Therefore I Am (More to Follow)' (2002) Derrida links the question of *our* animality to the issue of the cruelty of producing and killing animals as 'meat' on an industrial scale. The close link he sees between the treatment of animals 'in the past two centuries' and what has come to be considered the greatest evil among humans – genocide – testifies to his daring, for to make a comparison between the suffering of the holocaust victims under Nazi rule and that of animals under the rule of modern man was bound to attract criticism:

> As if, for example, instead of throwing people into ovens or gas chambers (let's say Nazi) doctors and geneticists had decided to organize the overproduction and overgeneration of Jews, gypsies and homosexuals by means of artificial insemination, so that, being more numerous and better fed, they could be destined in always increasing numbers for the same hell, that of the imposition of genetic experimentation or extermination by gas or by fire. In the same abbatoirs ...
>
> (Derrida 2002: p. 395)

What is significant in this thought experiment, which is deeply blasphemous from a moral point of view (but then morality itself is an element of the hegemony of rational man) is that Derrida sees a direct link between the pain and shame we feel in imagining such things, the opening up of compassion, and the emergence of its interpretation in 'law, ethics and politics.' Thus he understands the global movement for animal rights, in the light of the growing global 'exploitation of non-human-animals' as a necessary war following the emergence not only of industrial cruelty over against humans and non-humans alike in 'the past two centuries,' but also the discovery of our cruelty and of our potential for shame and guilt.

> War is waged over the matter of pity ... it is passing through a critical phase. We are passing through that phase and it passes through us. To think the war we find ourselves waging is not only a duty, a responsibility, an obligation, it is also a necessity, a constraint that, like it or not ... everyone is held to ... The animal looks at us, and we are naked before it. Thinking perhaps begins here.
>
> (Derrida 2002: p. 397)

In the citation above Derrida plays on the concept of nakedness as introduced by Levinas as a marker of the conception of ethical responsibility. There is also an allusion at work here to the biblical story of Adam and Eve being expelled from paradise after discovering sin, and its twin – guilt. When we imagine ourselves in the condition of being looked at, being under a silent accusation of committing a crime, we find that we have to answer by writing the law. The law is transcendent, for Levinas as well as for Derrida, and in writing it, we are aiming at it, aiming at justice, at what is right. That we can only aim at justice and that we will, by necessity, always fail to achieve it, is shown in the fact that we need politics (a game of confrontation, debate and power play) and ethics (the realization of our failure in reaching for the good, the infinite and indefinite guilt under which we live) to cover up the fact that law alone can never guarantee for justice to come to pass.

While Derrida describes in his essay on the animal the context of the war for/on animal rights, and for/on animality and the direction of 'thinking,' he also clarifies of what elements the deconstruction of the human–animal divide essentially consists. When we begin to think, that is to think ethically, we will realize the following:

1   The divide ('rupture') doesn't define two clearly separated domains – of 'human' and 'animal.'
2   The multiple and heterogeneous border of this divide has a history (the autobiographical history of anthropocentrist subjectivity), and should be traced as such.
3   Beyond the human side (which is heterogeneously delineated) there is not one category, 'Animal,' but a 'multiplicity of organizations of relations between living and dead' (organic and inorganic) (cf. Derrida 2002: 399).

Playing, as always, with language, Derrida insists (Derrida 2002: p. 415) that we use the word 'animot' instead of animal, to indicate the multiplicity of life forms that we have subsumed under one word. In our use of words we have, he says, also deprived the animals of 'the word.' In the end, confining his thought mainly to criticism, Derrida does not seem confident that we could 'give speech back' to non-human life forms, as the anthropologist-become-philosopher Kohn tries to do, by building a semiotic bridge to cross the human–animal divide (in his reconstruction of the worldview of peoples of the Amazon). Of course, even Kohn does not believe in what Derrida calls a 'homogeneous continuity between what calls itself man and what he calls the animal' (Derrida 2002: p. 398). He does trust, however, that we can communicate, bridge the divide, even while recognizing that it may not be desirable, or in any way possible, to erase it.

Enlightenment racism

Derrida's work is always an attempt to offer therapy to a thing called 'modern culture' or, alternatively, 'modern thinking' or 'industrial society' or 'Western modernity.' This therapy focuses on the root cause of its ailing condition: philosophy as onto-theology, that is, in essence a mythology about man's greatness, his unique role in creation bestowed on him by his own all-mighty God, which has clothed itself as reason, and in modern times even as science. This 'white mythology' is clear in the Kant quotation in the first section of this chapter, that declares anthropology to bring knowledge of the world. That we can know the world alone through knowing man expresses the arrogance of modern thinking. The therapy Derrida offers, consists of a form of philosophical psychoanalysis, bringing to light what modernity, as a culture of reason, keeps hidden in its shadows. In these shadows we find, essentially interlinked with the question of anthropology, and of the animal, the intrinsic racism of modernity. Racism and speciesism are, namely, of the same root, and function in similar ways. Not only are the animals not all one group, but those who call themselves (real) humans also segregate themselves from others like themselves they see as not completely human (Untermenschen), as of a different (othered) race, not worthy of full human dignity. Even today, racists repeat the same imagery tirelessly, calling their targets animals, monkeys, pigs or cockroaches. Not only from the mass killings engineered by the Nazis in World War II, but also from the Rwanda genocide in 1994, we learned that the actual killers are roused to do their job by the planners, the murderers in *sensu stricto*, through talk of 'vermin.' The people who are to be killed are not just identified as more animalistic than those of the 'master race,' but even as, supposedly, the lowest of the animals – vermin: those animals that are able to destroy 'human' culture, its crops, the basis of everything else cultural.

A philosopher who has tried to do some of the work that Derrida envisioned to deconstruct the human–animal divide – the tracing of the history of this multiple and heterogeneous border – is Emmanuel Eze (1963–2007). In the reader he composed of classical European texts on race, published with a commentary under the title *Race and the Enlightenment* (1997), he has shown, by means of textual analysis, that the heritage of modern philosophy – the Kantian and Hegelian construction of the idea of humanity as the center of 'our' (the Western) understanding of the world ('all philosophy is anthropology') – was built on the simultaneous construction of another 'other,' next to the animal, a not-quite-human: the 'savage,' the black man. This other was not supposed to have a culture of his own, let alone a political or legal system. In his book *Achieving our Humanity* (2001), Eze develops his own systematic discussion of the question of how to overcome racism in philosophy. It contains an especially interesting interpretation of the work of Kant, who is seen considered to have provided the philosophical foundation for what counts as valid judgment in modern science as well as in practical philosophy. At the core of his work was a normative view of what humanity should be, which Kant himself called his 'pragmatic anthropology.' It contains the comment that

The human being is destined by reason to live in a society with human beings and in it to cultivate himself, and to moralize himself by means of the arts and the sciences. No matter how great his animal tendency may be to give himself over passively to the impulses of comfort and good living ... he is still destined to make himself worthy of humanity by actively struggling with the obstacles that cling to him because of the crudity of his nature.

(Kant 2006 [1798]: pp. 229–30)

Here we see how the human–animal divide in Kant functions differently than Descartes's theoretical distinction between spirit/mind and matter. That mind-matter dualism just referred to two substantial categories, that divide all of reality between them forever, so to speak. With Kant the animalistic side of the human being becomes a condition of passivity which the human mind has to *overcome*. Thus the demarcation of what is supposed to be truly human has become *practical* – described as a condition we should struggle for, so that we can *make* ourselves human.

The struggle for humanity is not done by every individual for itself, though. The free, rational, human being will, according to Kant, subjugate itself to the public interest, and thus to 'civil constraint':

... in so doing they subjugate themselves only according to laws they themselves have given, and they feel themselves ennobled by this consciousness; namely of belonging to a species that is suited to the vocation of the human being, as reason represents it to him in the ideal.

(Kant 2006 [1798]: p. 234)

What Eze now aims to show is that, as beautiful and idealistic as this central principle of Kant might appear, it does not extend to all human beings as we know them. In the main text of his *Anthropology from a Pragmatic Point of View* (2006 [1798]), Kant describes theoretically,[2] to what extent the human races differ. This leads Eze to the conclusion that

... the lives of so-called savages were governed by caprice, instinct, and violence rather than law [which] left no room for Kant to imagine between the Europeans and the natives a system of international relations, established on the basis of equality and respect.

(Eze 2001: p. 78)

This means that the peoples who would be subject to colonization were not considered to understand law, let alone to be able to subjugate themselves freely to it. Therefore the Europeans, who were in the possession of such understanding and discipline, were allowed to bring them under the law *by force*, legalizing colonial rule on the principle of humanity.

Georg Wilhelm Hegel brought Enlightenment racism even a step further, by denying 'the black man' a place in the vehicle of humanity proper: history. 'The negro is an example of animal man in all its savagery and lawlessness … we cannot properly feel ourselves into his nature, no more than into a dog' (cited in Eze 2001: p. 24). He denied black human beings not only the understanding and factual institution of law, but also a place in history. 'Africa' represented for him 'nature' as opposed to history. When we think that history for Hegel essentially meant progress, we can understand that he saw the subjugation of Africans under colonial rule not as just a necessity (like Kant did), but rather as a gift to them, bringing the opportunity to these 'natural men' to enter history and to partake in the 'progress of mankind.' Of course, it was not only philosophers who formulated such thoughts: they were just the most eloquent, perhaps, in the networks of theologians, biologists, historians, missionaries, politicians, businessmen, white ladies doing 'good works' and others, who together managed to help bring forth what Derrida called the *history of animality*, as it was applied to those belonging to non-European/white nations.

That this history is also a history of what, according to Derrida, 'we call thinking' has recently been provided with historiographic proof by Peter Park, who investigated how, in the same period of time that the abovementioned thinkers were publishing their major works (1780–1830), racism constituted one of the central elements of 'the formation of the philosophical canon.' By digging into the now often forgotten textbooks which were to decide for the following two centuries (the same period that Derrida demarcated as the centuries of the invention of modern animal cruelty) what would count as 'true philosophy,' Park has shown that

> … the exclusion of Africa and Asia from histories of philosophy is relatively recent. It was no earlier than the 1780s that historians of philosophy began to deny that African and Asian peoples were philosophical.
>
> (Park 2013: p. 1)

This denial, not coincidentally, was introduced at the same time that religion and philosophy were ever more strongly segregated, which made it possible to claim that Africans and Asians had religion, but not philosophy. At the root of the construction of the modern 'history' of philosophy, Park sees the Kantian idea of taking a normative principle to decide what belonged to philosophy and what did not (to decide on what constituted valid philosophical reasoning).

> If Kant's philosophy provided a definition of philosophy and principles by which the history of philosophy could now be organized, it also provided principles of exclusion.
>
> (Park 2013: p. 7)

What should make us think about what is considered to be 'thinking' is the fact that this Kantian 'revolution,' which provided a normative principle for the demarcation of philosophy, is generally seen as having made the history of philosophy into a more 'scientific' field than it used to be: in fact, this meant that phenomena as they appear to us, the empirical experiences of actual life, were banned from its domain. So also the phenomenon of those non-European people's thinking was banned, which became all the easier since they were considered by certain philosophers to be not fully human.

The above makes it understandable that it can hardly be a coincidence that many authors who interpret the spiritual understanding of human–animal relations as it can be found in shamanistic cultures philosophically, such as Bamana (2014) and Kohn (2013), are not philosophers by training, but anthropologists. The relative distance which they have to the straight-jacket of the philosophical canon, as it is still taught in the present day, helps them to stretch Western philosophical concepts such as 'meaning,' 'thinking' and 'ontology' to refer to dogs, trees and other 'animot' (also to creatures of the more vegetative kind). A philosopher by training, like Derrida, obviously takes a more cautious road. He is unable to decolonize thinking in a concrete manner, or make the traditional philosophical concepts more inclusive. Instead he deconstructs them, to first offer therapy to (modern Western) philosophy itself, in order to enable it to overcome its sickness of exclusion. A therapy which was offered from a rather different position, by a thinker who was also a colonial subject and who was a psychiatrist by training; a therapy that was written for the black thinker, Frantz Fanon's works have also served to awaken many white, Western thinkers from their racist/speciesist slumber. Let us see what his approach adds, in the sense of a criticism of the biologizing effects of racial categories, but mostly as a design for a philosophy which thinks about the future of humanity.

Psychoanalysis for white philosophy

In *Black Skin, White Masks* (2008 [1952]) the young psychiatrist Frantz Fanon (1925–1961) testifies to the difficulties faced by a colonial subject, a black man moving to the 'center of the world' (Europe) in affirming himself as a man and as a human being. Arrived at this center, the gaze of the other, making him black, confined in his skin, empties him out before he can speak. This revolutionary book is not a typical philosophical discourse, building a thesis on assumptions and by means of argumentation. *Black Skin, White Masks* (Fanon 2008 [1952]) is written in a form that expresses its aim: to think not from general concepts, but from failures. This is the approach of psychoanalysis, and not that of what is commonly seen as philosophy. All the same, in its criticism of philosophy as a white man's instrument for ruling the world, it provides a thinking approach – an understanding of the world, but from a different point of departure than the usual one: 'not from a universal viewpoint but as it is experienced by individual consciousnesses' (Fanon 2008 [1952]: p. 123).

We could point to obvious similarities between Fanon and Derrida, who both, although in very different conditions, undertook the journey from a colonial space (Algeria, Martinique) to the center of power, Paris, to pursue further studies. Whereas the Jew Derrida eventually was expelled from his native land as representing the white colonial class, he was considered 'very black and Arab' in France, as not completely fitting in. Fanon describes a quite different experience of coming to the 'mother country' however, one of being reduced to a racial category – a reduction that empties him out completely, leaving only the outside, the color of his skin. When on the train a child calls 'Mama, see the Negro! I'm frightened!' Fanon comments:

> ... the corporeal schema crumbled, its place taken by a racial epidermal schema ... I was battered down by tom-toms, cannibalism, intellectual deficiency, fetishism, racial defects, slave-ships ... On that day, completely dislocated, unable to be abroad with the other, the white man, who unmercifully imprisoned me, I took myself far off from my own presence, far indeed, and made myself an object.
>
> (Fanon 2008 [1952]: pp. 84–5)

Thus Fanon shows the effect of the 'idea of humanity,' which is supposed to be abstract and universal but which through its alliance with denigrating ideas about 'negroes' in fact excludes human beings from the public sphere of reason (the Kantian sphere of humanity) and imprisons them in the color of their skin. 'The Negro is an animal, the Negro is bad, the Negro is mean ... Mama, the nigger's going to eat me up' (Fanon 2008 [1952]: p. 86).

To rescue his black reader from the objectification and dehumanization that even the social sciences do to her/him, Fanon does not stop at being just critical of 'white' thought, but undertakes to articulate the humanity of the black person. This should not, however, be done by integrating himself into the European idea of a supposedly non-race-sensitive humanity. This 'Hellenistic' idea of humanity, says Fanon, considers black persons to be like animals (Fanon 2008 [1952]: 127), biologizing and sexualizing them, while desexualizing the white man as the embodiment of universal reason. Like Derrida, he links the question of the human–animal connection to politics and ethics, albeit in their suppression: when the black man can be seen as the biological animal-human, the white oppressor does not have to see his own violence toward him. In order to evade the dangerous situation this 'humanity' in fact creates, *Black Skin, White Masks* (Fanon 2008 [1952]) seeks to submerge itself into the shadow side of white culture, and to investigate the fake aspect of the 'animality' projected on the individual of African descent. Thus, although Fanon has the black individual in mind for his liberative project, implicitly he also criticizes, pre-figuring the Derridian approach, the idea of animality in general (being described in terms of predatory behavior, being driven by sexual urges, et cetera) as an artificial biological objectification. Using the analytic tools of psychoanalysis, he shows that the collective guilt of '...white society – which is based on myths of progress, civilization, liberalism, education, enlightenment, refinement...' (Fanon 2008 [1952]: p. 150) is transferred to the 'scapegoat': the black man. Fanon's aim is not to liberate white society from its guilt, nor to show its sensitivities as expressing a possible openness for an awareness of justice, as Derrida does. This is not because he is writing in a time when decolonization was still underway, but because he is not interested in the problems of the oppressor culture. Fanon hopes for his book to '...be a mirror with a progressive infrastructure, in which it will be possible to discern the Negro on the road to disalienation' (Fanon 2008 [1952]: pp. 141–2). This means that a truly human being does not have the past in mind in his actions, founding his actions upon nostalgia or resentment, but rather the future and the justice and freedom that are his own. His aim is to become 'actional' rather than 'reactional' (Fanon 2008 [1952]: p. 173). In not wanting to repeat the injustices perpetrated by white society, Fanon calls on his reader to find new ways to envision humanity, and human society.

In addition to being a psychoanalyst and a writer, Fanon also contributed to the Algerian freedom struggle by training rebel fighters, and so he was often interpreted by readers within the context of white cultures as promoting violence. Here we can see the double standard at work of what counts as violence. While European countries' violent interventions in other parts of the world are sold as rescuing, peace-keeping or stabilizing operations, those who use violence to throw an oppressor army out, are named rebels or terrorists. For Fanon it is clear: violence should not be reactionary, for it will then always repeat what it is up against. It shall only be *for* the freedom to be human. Although this psychoanalytical approach of Fanon aims to liberate the black reader from mental oppression, it also helps to liberate (as an undeserved kindness) white culture, especially white philosophy – not from its guilt, but from its lack of awareness regarding its own violence toward those it has called 'almost animals.'

In the end, we may conclude that taken together, the decolonizing and the deconstructing approach, work in concert for the preconditions that are needed for the negotiations we hope for. First, they create options for those who have been colonized and animalized to appropriate the position of thinkers and selves, of subjects and gazers, exposing those who call themselves civilized and masters and making an end to their reign. Second, and simultaneously, they create awareness for the shadows of Enlightenment/modern philosophy – for its support for the cultural and physical genocides of those it named others; its masking and legitimation of their enslavement and dehumanization; and, in the shadows of those shadows, the moral condoning of the use and abuse of 'animot.' Third, they expose the silencing of the voices of those 'on the other side' of (white) humanity, the 'savages.' And, further along the blurred lines of the human–animal divide: the silencing of the thinking forests, the wild animals who might still offer guidance if asked politely, even the 'domesticated' intimate strangers living with us – they all have been colonized too, under the cover of the civilization of 'humanity.'

Notes

1 This last-mentioned struggle has been described and put into historical context in an attempt to write from the viewpoint of the Lakota whom it concerns in Ekberzade 2018.

2 Theoretical is here opposed to pragmatical, in other places in Kant to practical. In present day discourse we would call the theoretical the empirical – that which is not subject to willing, but only to description and explanation.

Chapter 6

Vital force

A Belgico-African missionary's spirited philosophy

Bantu philosophy, culturalism and its critique

In this chapter we will look into the work of the Belgian missionary Placide Tempels (1906–1977) who published, in 1946, a small book titled *Bantu Philosophy* (1959). Although the title says it treats of philosophy, in fact it intervenes in several disciplines as they were in Tempels's time. The way this was done is of central interest for our project, as I will point out. *Bantu Philosophy* (1959 [1946]) intervenes in anthropology or ethnography, as Tempels calls it, because he objects to its uncritical stance. It intervenes in philosophy, criticizing its singular focus on Western thought, and opening it up to an African approach. Finally, it intervenes in theology, opening up a hermeneutical dialogue between traditional African religion and modern Christianity on the workings of divine grace in human life. Adding to that *Bantu Philosophy* (Tempels 1959 [1946]) does all this from the frame of a critical analysis of industrialized culture, and as such takes its place among works of critical theorists in sociology. The work is important to the present endeavor for the following reasons: first it provides an interesting attempt to ontologize cultural anthropology long before its ontological turn. It is of importance to ask whether Tempels takes a different approach than Kohn and Viveiros de Castro, and if so with which consequences. Second it also culturalizes ontology, a move which we have to investigate for its potential to create a field for negotiations concerning the worlds of different peoples. Besides that we should look at Tempels's attempt to outline the logic of a shamanistic way to relate to the world – and at the extent to which it is founded on dialogical exchange with the Baluba people with whom Tempels lived.

Together with the colonials, the entrepreneurs, the military and the governing class, came the ethnographers and missionaries to the colonized lands. Their goal was to understand and 'civilize' the 'natives.' The discipline of ethnography through which Europeans in times of colonialism acquired knowledge of the peoples they encountered, did so through the gaze that produced their othering. This othering took place by making use of a concept of ethnos, people, or of culture, life form, that was meant to explain in what these peoples could be distinguished from Europeans. What made them different. What was essential to their way of life. The effect of the anthropological gaze still rules over many postcolonial situations. The *culturalism* adopted by the ethnographers has often been adopted by the liberators of colonized peoples as well, claiming pure traditional cultures, untainted by the moral and psychological rot that colonial oppression created. Boele van Hensbroek describes the effects of this orientation, that goes together with the politics of identity:

> Culturalism, in this view, involves the combination of an essentialist idea of cultures (culture conceived of as a kind of 'body' or 'entity' held together by a strong internal coherence of even a core or essence) and the view that persons always 'belong' to a culture, that a person should be anchored in a culture, that the cultural framework is a condition for being authentic.
>
> (Boele van Hensbroek 2001: p. 136)

This view of cultures (plural) as coherent bodies or entities was dominant in cultural anthropology until recently, as well as in the young field of comparative philosophy. Intercultural philosopher Pius Mosima argues convincingly against this understanding of cultures, and claims instead that human beings

> ...in their daily lives have several overlapping 'cultural orientations,' which co-exist and from which they learn daily, and not just one 'culture' that combines claims of totality, integration, and boundedness.[1]
>
> (Mosima 2016: p. 23)

The danger of a culturalist approach for emancipatory movements consists in closing society off from the future, in a bid to find connection and meaning in the past.

> Such movements however are at risk to deny the historicity (change, life) of culture, as well as the need for new ways to live in a new situation. As much as the search for a golden age in precolonial times can be a necessary moment in the process of cultural decolonization, it may lead to ignoring the future oriented aspects of culture building.
>
> (Boele van Hensbroek 2001: p. 136)

An alternative use of the concept of culture is found in the work of Amilcar Cabral, leader of the independence struggle of Guinea Bissau, who wrote extensively on the multiple aspects of liberation, and defined culture in a dynamic, future-oriented way. He wrote that '...no culture is a perfect, finished whole. Culture, like history, is necessarily an expanding and developing phenomenon' (Cabral 2007 [1979]: p. 179). To Cabral culture is the creation of society, of people forming a social sphere, the congealed form, as it were, of its coming to terms with its internal conflicts. The consequence of this view is that culture is not essentially bound to geography, ethnicity or race, even though it is situated, arising in a certain place and time: 'The co-ordinates of culture, like those of any developing phenomenon, vary in space and time, whether they be material (physical) or human (biological and social)' (Cabral 2007 [1979]: p. 179). This dynamic understanding of culture is a powerful element in Cabral's revolutionary theory – theory that is directed to change. Opposed to the descriptive and explanatory understanding of cultures, that may function in a political order in which the group in power others the group that is manipulated and used, the revolutionary concept of culture is meant to stimulate people to negotiate ways to live together that overcome tensions within a society. Within the context of this work, I propose to put this future-oriented Cabralian concept of culture to new use: to articulate conditions for intercultural negotiations of the environment. Such negotiations have to be future-oriented, as they not only take the actual survival and well-being of future generations into account, but also their ways of being in the world, as well as their cultures – in their overlapping and contesting aspects. Thus the theorizing of Cabral, that was aimed at the forming of new nations, is transferred to our context of global environmental threats in a time of increased connectivity and contact between ways to be in the world. Environmental problems pressurize peoples to give more attention to their ways to relate to nature.

A romanticized longing for a more sustainable life – as we can see in many movements such as the tiny house movement, the completely self-reliant village and others – is simply not enough anymore. In this context where peoples have to learn from each other in order to keep the environment in shape, cultural anthropology plays an ambiguous role. Whereas most anthropologists strive toward a decolonization of their own profession, they often fail to do so in important points. They do try to give a voice to the ways of being in the world of the indigenous peoples they investigate, while these are struggling to adapt to the destruction of their ancient ways of life. They do present them as agents of their own fate, in alliance or in conflict with the multiple kinds of spirits that shape life (cf. Pedersen 2011). Still in their work, as I showed in Chapter 2, two things often lack. One thing lacking is a truly dialogical approach. If researchers do not submerge themselves into the lives of the colonized peoples, while simultaneously realizing they will always still represent the colonizer to them, as sympathetic and 'decolonial' as they may posit themselves, their frame of investigation will always still dominate the voices of the peoples they hope to translate to their Western academic audiences. The second thing lacking is the realization that no dialogue can take place unless those who have 'empire' written across their foreheads, don't accept to stand for the trial of the indigenous peoples – who should decide if they are acceptable as ambassadors without endorsement, or if they will be sent back to wait until their political consciousness will have grown and things be done to change the political relations between them. To take these two preconditions together in a brief conclusion: no academic engagement with indigenous peoples can measure up to what is at stake today if its results are academic products only, such as books, careers, prizes and speeches.

Within the context sketched here, it is of importance to re-interpret the lives and works of those who are now stored in what is called the 'colonial archive.' The case of colonial history, from both sides – that of the colonizers and the colonized – will have to be reopened, to find the actual struggles of oppression and resistance, the actual acts of collaboration and of sabotage – acts that can be found in individuals and groups on both sides of this war. This reopening is not geared toward finding truth after all or even to doing justice. It aims for creative instruments to invent humanity once more. Against this background, this chapter will return to the work of Placide Tempels, whose work happens to have started a long debate on the nature and status of African philosophy (cf. Mudimbe 1988 and Mosima 2016). The meaning of Tempels's work for African philosophy is, however, not the primary interest here, but rather the ways in which it tried to include spirited ontologies in philosophy, a move that reminds us of the recent ontological turn in anthropology. Tempels reproached ethnography for leaving out the critical approach to culture that philosophy can bring. Adding to that he criticized the colonial enterprise itself for its focus on making humanity into a community of consumers and producers of material goods. All the same, he avoided to lapse into a romanticizing of 'ancient' ways to be in the world as well. Finally it can be argued that he transformed the idea of philosophy itself to include the wisdom/knowledge present in 'magical' (shamanic) practices – thus clearing ground for a truly intercultural dialogue on what it is to be human.

Tempels's transformation: becoming Bantu

The life and work of Placide Tempels deserves renewed study, now the world of which he was a part, that of the Catholic mission by Europeans in non-European lands has come to an end. There are different reasons to take up such a study. His *Bantu Philosophy* (1959 [1946]), although written for missionary purposes, has had its largest effect in African academic philosophy – in the decades-long debate on the question whether there exists a specifically African philosophy, and if so, what are its characteristics. Tempels's book was one of the stumbling blocks that led African philosophers to take on this debate, as it had aimed to reconstruct a truly African, 'Bantu' logical system of thought, that could explain all the aspects of ideas and practices of the peoples with which Tempels had become acquainted in the Belgian Congo before independence.

Bantu Philosophy (Tempels 1959), published first in 1945 in Elizabethville, Belgian Congo,[2] has been discussed and critiqued for over 70 years now. Valentin Mudimbe, who provided an extensive account of the decades of discussion it raised among African philosophers, situates it 'Within the arrogant framework of a Belgian colonial conquest meant to last for centuries...' (Mudimbe 1988: p. 136). From hindsight, of course, World War II meant the beginning of the end of Europe's colonial empires. And *Bantu Philosophy* (Tempels 1959 [1946]), which soon was translated from the Dutch into French, German and English, was written just around that time – right at the point of transition when empires still believed arrogantly in their unending might, while their ideological and philosophical foundations were already crumbling. Its author, addressing the 'good colonials' didn't question the purpose of his missionary involvement – to civilize and to educate 'primitive man.' Still he changed the idea of civilization to include the wisdom of his African interlocutors – the Baluba people. In an attempt to articulate this wisdom, and to give voice to the reality it reflected, he reconstructed their understanding of reality as 'Bantu ontology.' Tempels wanted to intervene in a missionary and civilizing project that, he found, had failed, as it didn't really engage with people, with their values and commitments, and so could only create 'évolués.' Or to put it bluntly: black colonial servants who had adopted a European lifestyle, while struggling with existential anguish as they lost their true spiritual commitments, without having adopted others.

> We have now a multitude of 'evolved' persons, who look down with condescension upon their own race, but don't know how to live any more as they don't find meaning for it. The European thought and aspirations have been served to them in a way that makes them indigestible.
>
> (Tempels 1946: pp. 108–9, translation A. R.)

By reconstructing the Bantu philosophy, Tempels not only wanted to convince the 'good colonials' to adopt a more open mindset in dealing with African people; even more he hoped to give the Africans an instrument to take their ideas and practices, their entire life form, along while opening themselves up to the new culture and religion of the Europeans. Only if they would have the possession over the logical system inherent in their way to deal with the world, would they be able to they deal with the necessity (given the colonial context) to fuse two worlds reflectively. Their wisdom, their philosophy, Tempels held, was as strong as the European one, and this was the cause that they cannot leave it for another, not even if they wished to:

> Here stands one wisdom over against an other one, one way to under-
> stand the world over against an other one, one ideal of life over against
> an other one.
>
>   (Tempels 1946: p. 104, translation A. R.)

He doesn't make it explicit in this book, but it is clear from other sources
that the missionary had increasingly become convinced of the worth of the
values of his African interlocutors. For instance when he writes about

> ... their very elevated understanding of justice and social order, of good
> and evil, and ... the stubbornness with which the Bantu defend their
> right in what we call palavers, but what to them above all means restor-
> ation of life according to rank of life as willed by God.
>
>   (Tempels 1946: p. 115, translation A. R.)

Several authors have remarked that Tempels's attitude toward the Africans
he came to convert changed in the course of his life. While starting out as
a hard-core missionary who came to bring white, European values, he
increasingly turned into a defender of an Africanized Christianity and a
critic of the European life style and its condescending attitude toward
Africans. Since the 1970s, however, there has hardly been a return to the
sources on Tempels's life, even as so much has changed in theoretical per-
spectives toward the colonial archive. One of the scarce newer articles on
Tempels, by Deacon (2002 [1998]), which mostly relies on secondary
sources, makes a point for seeing Tempels as a victim of colonialism. After
the publication of *Bantu Philosophy* (Tempels 1959 [1946]), de Hemptinne,
the bishop of Elizabethville in the Congo tried to thwart Tempels, by
ordering him to return to Belgium, and by attempting to get the book
outlawed by the church authorities in Rome. Time was however on Tempels's side, and after a few years he could return to the Congo, to pursue
his desire to work for an Africanized Christianity, and in this frame he
later founded the Jamaa – a religious movement that centered family life
as the core of Christian life.

From the beginning, Tempels seemed to have been destined to *meeting*
Congolese men and women, instead of only educating them. When he was
assigned to do missionary work in remote villages, namely, he didn't follow
the advisory plan for the layout of a missionary post:

> On arriving in the appointed village, the missionary worker inevitably
> builds a homestead. The home would in most cases, be constructed
> outside, and some distance from the village. The local dialect would be
> learnt, while the missionary would recruit a few 'disciples' for assistance
> in building the structures of a church and a school.
>
>   (Deacon 2002 [1998]: p. 122)

The idea behind this approach must have been to not let the 'pure' religion as it was brought get tainted by village life, and to lead the indigenous people symbolically out of their traditional village into the new faith. The effect of this approach must have been that the converts experienced an emotional distance toward the new religion – something that Tempels deplored. He wanted to do things differently, therefore

> ... Tempels not only resided with the traditional people in their villages, [he] ... related to his converts from within their cultural circumstances, and functioned in not rendering a cleavage between the people and the foundations which rendered them their ultimate meaning.
>
> (Deacon 2002 [1998]: p. 122)

The core idea of the Bantu ontology that Tempels constructed out of his encounters with his neighbors is often rendered as vital force, a primary reality that can be opposed to the primary reality in Western ontology – static being (things). In most representations of *Bantu Philosophy* (1959 [1946]) this opposition is stressed to the point that it becomes a purely philosophical opposition of schemes of ideas. Below I will discuss how changes that have been made to the text in the French and English translations may account for a simplistic rendering of the philosophical matter. For our purpose here it suffices to state that life to the Bantu was, according to Tempels, all about trying to increase vitality and fighting its decrease. Similar to the Amazon ontology, as deduced from elements in the Runa way of life by Kohn, the Bantu perception of reality presupposes a non-dualistic world, meaning a world that doesn't oppose mind against matter, and which takes everything as living essence or power. Tempels goes beyond what Harvey named 'old animism,' as he doesn't claim that everything 'has a soul' or spirit, but that there is only individualized power, or maybe we would say today – energy. It is in the exchange on this energy level that what the West has called 'magic' can be understood.

Tempels approach arose from a frustration over what ethnologists as well as missionaries had left undone in their dealings with their African subjects: to ask them about the meaning of their practices and beliefs – an approach which he now introduced. In a biographical note written around the time that *Bantu Philosophy* (1959 [1946]) was published, Tempels described the experience of learning from his Luba friends as follows:

> So, I would look at this person and ask him: 'What is the matter with yourself? What is wrong with him? What sort of person are you? What are you thinking? What do you want the most? What about your magic medicine? What does it mean? How does it work?'
>
> (Tempels 1962: p. 37)

Tempels at first wanted to take the position of the 'Bantu,' to understand his perspective in context, hoping he could then return to his Western educative goal with more background. Instead something else happened. He was elated to find his conversation partner able 'to express himself clearly, also discovering himself for the first time through reflection. [He found] the depths of his personality, the mystery of his being and of his soul' (Tempels 1962: p. 37). Even though his surprise may have sprung from an attitude of condescendence, the text goes further to state that not only the African man (who still has no name), but Tempels too found himself through reflection in that same conversation:

> What joy, new to the both of us to discover we resembled each other and, what is more, to see we began to 'meet' each other soul to soul. And there I had been thinking that after having discovered the Bantu personality, I could have gone back to being the pastor, the governor, the doctor. Even though I mastered a technique of appropriate language use to 'teach' Christianity, I suddenly realised that in this man to man meeting and soul to soul encounter from one being to another, we had evolved from mutual acquaintance to getting on well, and finally, to love ... and [I saw] that precisely Christianity had just been born and had already begun.
>
> (Tempels 1962: p. 38)

In this experience something happened that left its mark on the rest of Tempels's life. A personal remark from the time when he returned (for health reasons) to Belgium, and he looks back at his time in the Congo, makes clear that somewhere during his exchange with the people he met there, he committed himself to their understandings and systems of thought and practice:

> Personally, I have been lucky to have accomplished myself and to have blossomed out to the heart of my self thanks to the Bantu man, and now not to live anymore but answerable to him.
>
> (Tempels 1962: p. 40)

It is safe to say that somewhere along the way Tempels had been converted himself, to the way to relate to the world he had tried to capture reflectively in *Bantu Philosophy* (1959 [1946]). In that book we already see his criticism of the colonial enterprise as such – to turn humanity into a collective of producers and consumers, forgetting the needs of the human soul. These needs, he found, were better addressed in the Bantu wisdom, which is the main reason we should, in his view, take its ontology, and the concomitant psychology and moral thought, seriously as a contender of Christianity – with respect to what can nourish emotional and spiritual needs.

Industrialization, however, the introduction of an European economy, permanent raising of production – all that is not necessarily a measure of civilization. On the contrary, it may lead to the destruction of civilization, unless sufficient account is taken of man, of human personality.

(Tempels 1959 [1946]: p. 172)

In the attention Tempels gave, in the last chapter of his book, to the political-economic context lies another reason to revisit this text. Combined with his dialogues of the heart, his awareness of context seems to protect Tempels's work from being just another attempt to appropriate a non-Western ontology into the generalizing gaze of white academic theory.

From the 1930s on up till the 1960s, Catholic priests in Europe were very active in social work for factory workers – among them the youths who, often from the age of 12, were put to work on the production line. This movement reflected a then radical social theology, in which 'being with' the workers was the ideal which the priests aimed to live, thus going against the hierarchical structures that characterized the church. Tempels, in his part of the world, was close to this movement and its ideals. Jozef Cardijns, one of the frontmen of the movement in Belgium (later appointed a cardinal), was his friend, and later stood in for his work with the Jamaa movement. Like the social action priests, Tempels worked for several years in young mining towns. Now and then he was asked to speak, write an article or give advice on the basis of his 'old' work on Bantu philosophy. In 1957, when he was working in the mining town Musonoi, he apparently got such a request from the editors of a book called *Aspects de la culture noire* (*Aspects of Black Culture*). His answer gives a clue how at that time Tempels had gone beyond dialogue as the aim for intercultural exchange in the context of the colonial situation:

I do understand what you are asking. It is about, I believe, the dialogue of real-life Bantu ontology with Western culture. That would then be my field. Right, but not quite right.

It would indeed be possible to establish this dialogue, and I could write an essay on it ... If I had not been immersed in this crowd, which engulfs me and in which I am trying to fit in ad lib. So you see that, for me, it is not about starting up a dialogue between a Bantu point of view and Western culture. Neither is it even about starting up a dialogue between a European and the Bantu people. To me, it is about an entirely different enterprise. I have to immerse myself entirely in the mentality, psychology, even in the life of the Bantu man. I have to rid myself of everything Western so that I myself can become Bantu with the Bantu people. It is about accomplishing this communion with them, this union of life which they all long for. So there is more to it than [merely] a dialogue between Europeans and Bantu people. There is communion, union of life between them and me, only as far as their Bantu life is concerned. And in this communion, we look together at our entire life, together we re-examine our entire life, looking into its tendencies and fundamental aspirations ...

(Tempels 1958: pp. 172–3)

Here Tempels seems to have arrived at a point beyond dialogue. Dialogue as a goal may still raise suspicion, especially when it is initiated by someone who represents the oppressor. Tempels instead aimed to take the perspective of the other, and to look, with him, at life. This move can be seen as a preparatory step for the kind of negotiations this book is about – we need negotiations because a silent war is going on in which land, cultures and lives are being lost. Before the conditions of any dialogue can be negotiated, there can be no dialogue. And before those conditions can be negotiated, it may be a good example to annihilate the white perspective as much as possible, as Tempels tried to do, and submerge into the way of life of the other.

Plural ontologies

Tempels's attempts to found African Christianity on a logical system of African values and insights into reality was not accepted well by his Catholic superiors, at least not at the start. His work was also disputed among Catholic theologians for its phenomenological philosophical approach, that didn't sit well with orthodox Thomistic (scholastic) philosophy that was then in fashion among Catholic theologians. *Bantu Philosophy* (Tempels 1959 [1946]) more specifically was close to the – then – *new* current in Catholic philosophy: existential phenomenology. It centers the phenomenon: that which reveals itself in our experience. It attaches no relevance to the presupposition of an objective truth that would be knowable next to what can be experienced, as Thomism did. Phenomenology does not accept an argument on the basis of what is in the mind of God, or on the basis of pure reason. It works with situated experience – by human beings – of their world, and aims to clarify from within what reveals itself in such experience. Tempels was also criticized for his claim that there is truth in Bantu ontology next to the European ontology he was trained in. This claim presupposes that philosophy accepts the condition of culture to be an integral part of thought. Thomist philosophy on the contrary holds that a universal philosophy, the philosophia perennis, is accessible to the intellect before or beyond any cultural differences.

A certain father Boelaert, a missionary like Tempels, wrote a critique of *Bantu Philosophy* (1959 [1946]) on the basis of this position. His critique is especially interesting as it shows that the inclusion of 'magical' or 'spirit' phenomena in philosophy that Tempels holds possible presupposes that one has overcome the demands of Aristotelian dogma, one of the main pillars of Thomistic philosophy. Writing on the correct way to analyze the structure of being, Boelaert posits:

But all those 'compositions' are 'real' in a metaphysical sense; those constituents of being are no beings, but principles of being, ontological
postulates, that look upon the causes of being. The principles can be
separated 'secundum rationem, licet non secundum rem.'... The essence
and the existence 'non possunt separari etiam potentia absoluta Dei.'

(Boelaert 1946: n.p., translation A. R.)

What this very technical language says is that we can distinguish the form/
essence of a thing from its material substrate, but that 'in reality' the two are
intrinsically connected. Only the absolute power of God can divorce them
(so that the human soul may live after the human body dies). As we will see
in the next paragraph in more detail, Tempels uses the concept of 'essence'
(over against existence) throughout his book, apparently to articulate how
spiritual phenomena in the 'Bantu' system of thought are real. While now,
according to Boelaert form and matter (essence and existence) are never apart
in reality, and this is the tenet of the eternal philosophy revealed by means
of the pagan philosopher Aristotle to the Christian doctrinal teacher Thomas
Aquinas, consequently Tempels cannot hold that there is a 'Bantu' reality, in
which magical things happen, next to the universal reality as laid down in
Western philosophy. Boelaert concludes his critical review on a pragmatic
note, allowing Tempels to be right in seeing 'strengthening of life' as a
central desire in Bantu culture, and proposing it as a useful basis for the
mission to work from.

It is worthwhile to cite Tempels's reaction to Boelaert's critique more
extensively, as it goes right to the point of the plurality of ontologies that I
already discussed in Chapter 4, and which is of central importance for this
work. Tempels replies to Boelaert in a letter of 1947 to his friend Hulstaert

1 That I certainly did formulate a concept of being.
2 That this concept of being expresses an aspect of reality that differs
   from the static one which is predominant in the West.
3 That this ontology is not non-sense, as father Boelaert claims, but
   that it is also a Philosophia perennis, the eternal philosophy of forces
   that exist universally next to the static philosophy.

(Bontinck 1985: pp. 154–6, translation A. R.)

The static philosophy is here the classical Aristotelian metaphysics of substances to which qualities can be ascribed (accidentia). Increase of life,
strengthening of life – the movement about which Tempels writes, cannot,
according to Boelaert, be considered the core element of a specifically African
reality, as substance is the sole core of any reality. This somewhat technical
and scholastic discussion now makes it clear: Tempels holds it to be logically
possible to claim the truth of different ontologies as eternal philosophy, and
thus accepts their plurality. In his answer he also stresses the 'cultural'

viewpoint, which by necessity leads to the ontological plurality Tempels introduces. The upshot is, therefore, that the acquaintance with non-European peoples that Tempels acquired through his missionary work turned Aristotelian metaphysics as the only acceptable philosophical foundation for Christianity *obsolete*. To invite Bantu thought, wisdom and spirituality into an intercultural philosophy, Tempels deconstructed the pillars of Western hegemony in philosophy.[3] He turned the idea of conversion around, suggesting: what if not the metaphysical ideas, but communion and solidarity, make up the heart of Christianity? If that is the case, one can accommodate for the entire 'Bantu' system of ideas, as long as it is internally consistent, and as long as it undergirds the same practices of familial love and awe for God and his creation that Christianity aims for in the first place.[4]

Vital force and the Bantu ontology

As indicated above, the English and French translations of *Bantu Philosophy* (Tempels 1959 [1946]) are highly problematical, a fact that has gone unnoticed for a long time, as most researchers in African philosophy studied the work solely in its French or English versions.[5] In his 2006 dissertation on African philosophy, Dutch philosopher Henk Haenen declares he will refer to the French translation of 1948, as the book first found its way to its African readers in this version, meanwhile warning that one always has to return to the Dutch – '… returning to the original Dutch text is certainly of importance where the French translation in some cases exhibits salient differences' (Haenen 2006: p. 65n237, translation A. R.). These differences seem, where they are indicated by Haenen, to make the text less precise and more negative toward Africans. For example, an expression which reads in Dutch as 'a university-educated Black' has been rendered in French as 'un indigène' without specification. A short article from 1993 by Willem Storm,[6] on the translation matter already noted the racist turn the text often takes in its English translation (that was made from the French).

> There are a lot of small deviations in the text which are probably inherent in any translation…. Each of these smaller deviations is not very harmful, but it might not be mere chance that in none of these cases the Bantu benefit from the translator's deviations.
>
> (Storm 1993: p. 69)

The mentioned reference to 'a university-educated Black' becomes in English 'a Bantu.' In a similar vein 'languages' becomes 'dialects,' an 'animal' becomes 'a wild beast,' 'their point of view' becomes 'the subjective point of view of the Bantu.' What is more, depreciating remarks seem to be added, and appreciating ones by Tempels concerning the leadership qualities and rational potential of the 'Bantu' are left out in the translation (Storm 1993: pp. 69–70). Besides, the translations seem to ignore the specific articulations of Belgian Catholic philosophy of the twentieth century. For all these reasons I will provide my own translations of the Dutch.[7]

Apart from the racist language in the translations, which make Tempels's remarks on how he was influenced in his thought by his African interlocutors harder to grasp, we should be interested in translation issues concerning the core subject matter of *Bantu Philosophy*: in what is translated as its 'Bantu ontology' of 'vital force' (Tempels 1959 [1946]). Both translations are actually problematic. Many interpretation questions arise from the fact that Tempels was not a professional philosopher but a priest, and that in general he doesn't refer to other philosophers or argue for certain idiosyncratic ways in which he uses philosophical concepts. It is to be expected however that his philosophy professors in the seminary had been versed in existential phenomenology, which was then trending in the southern parts of the Netherlands and in the Dutch speaking part of Belgium.[8] This may explain that where the English gives the section heading 'Bantu ontology' the Dutch has 'the theory of essences of the Bantu.' Although Tempels sometimes uses the term 'ontology' too, it seems to be no sloppiness on his part that he mostly uses the word 'wezensleer' (theory of essences) instead of 'zijnsleer' (theory of being).

A thorough investigation of the kind of philosophy that was taught in the Belgian seminaries where Tempels studied, or of what he read afterwards, is still lacking. One may expect however that during his studies, from the late 1920s on, the phenomenology of Edmund Husserl may already have had a profound impact on the philosophy professors in the seminaries. In fact, the movement that came to be called 'neo-thomism,' and that prevailed in Catholic institutions during a large part of the twentieth century in the said region, aimed to save Thomist-Aristotelian philosophy by transforming it on a phenomenological and existentialist basis. The distinction between the theory of being and that of essences that seems to be important to Tempels, was already prominent in Hegel, but is taken up by Husserl in new ways, in his investigations into consciousness, intentionality and perception. Without going into this too deeply here, it is important to remark that Tempels throughout *Bantu Philosophy* preferably uses 'wezens' (essences) instead of 'zijnden' (beings) (1959 [1946]). The confusing thing is that 'wezens' in ordinary language can also mean 'creatures' or, indeed, 'beings' – which are both translations we find in the English version of the book. In more technical philosophical language the difference is meaningful. Although this should be corroborated by further research, I put here the hypothesis forward that Tempels made a technical distinction here on purpose. The reason being that the Husserlian account of essences could allow for the non-material active agents (spirits) or spiritual energies that are so typically 'African' to the Europeans who studied African ways to be in the world in the past century. In a recent article on the concept of essence, Zhok argues that Husserl gave it a special meaning, in order to 'save' the Aristotelian idea of forms, in a bid to counter the tendency of positivism to only account for matter and mind as real.

> The point is that events and states of affairs in the world emerge by taking shape and their form (their essence) establishes how they can affect and be affected. In this 'how' is included all we need to make room for the efficaciousness of essences.
>
> (Zhok 2012: p. 130)

The theory of essences in Husserl allows for an in-between with respect to mind and matter that might have appealed to Tempels, or to a philosopher in his environment who passed the idea on to him.

'Vital force' is also a problematic translation – of what Tempels calls alternatingly 'levenssterkte' and 'levenskracht.' 'Levenskracht' as it is used in ordinary language is translated simply with 'vitality.' This would make the Bantu doctrine of essences one of vitality, which sounds strange because of the combination of technical and ordinary language. Choosing a more technical translation, therefore, we will have to render 'levenssterkte' as 'strength of life.' 'Kracht' can be translated as force, power or potency. If it should also be a synonym of strength, we might best choose 'power' – which gives us then the expression 'power of life' as translation of 'levenskracht.' This can of course be taken as a synonymous expression of vital force. The point of a new translation is twofold, however. First it has to do justice to the Dutch as best as possible, and second, one has to wonder what Tempels wanted to confer as truly 'Bantu.' The first point can be checked by translating back – vital force would be best translated back as 'vitale kracht.' The second point follows suit – the concept of 'kracht' or 'force' standing on its own has strong semantic overtones stemming from the natural sciences, which makes us interpret Bantu ontology as a dynamistic system of forces, which is exactly how Tempels has been read. If we choose a completely alternative rendering of 'ontology of vital force,' namely the more correct 'doctrine of essences of strength of life,' we get a different conglomerate of meanings. One that doesn't recognize beings by the share of vital force they express, but rather essences (or 'forms'), that vary with regard to the quality of the strength of life that, so to speak, flows in them.

What is important here is whether Tempels may have done a good job in his attempt to give a voice to the Baluba, who had a system of thought, he concluded, but who had never written this down or even articulated it. Again, my issue is not whether *Bantu Philosophy* (Tempels 1959 [1946]) gave a true representation of the thought of the people he lived with, or of African peoples in general. What interests me is whether the way he tried to give them a voice did a good job to open up potential negotiations between cultures/peoples/groups with conflicting relations to nature. From this perspective we should turn our attention away from the question whether Tempels reduces Bantu ways to be in the world to Western philosophical concepts – which he did, obviously. He translated, to open up the Bantu world to Europeans and to the 'Bantu' people alike – he opened it up to them differently than it already was, in academic philosophical discourse. And he opened it up to both in order to *negotiate* both worlds – that of the Bantu and that of the European – as far as they can be shared. This is confirmed by the fact that it is of importance to Tempels to treat both systems of thought as epistemologically equal:

> The insight of the primitives in the nature of things, as well as the far-fetched distinctions of learned professionals, are intellectual knowledge of essence, and both are not essentially different as knowledge. Both are knowledge of essence, metaphysics, and the system of thought that is built on a specific concept of essence, is philosophy.[9]
>
> (Tempels 1946: p. 24, translation A. R.)

Similarly he stresses that he expects that the Bantu will confirm, after the work of constructing the Bantu ontology will be done, that their knowledge systems now can communicate on a par:

> They will recognize ourselves in our words, and answer: You have heard us, you are completely familiar with us, you 'know' like we 'know.'
>
> (Tempels 1946: p. 12, translation A. R.)

Now the question is *what* Tempels opened up in his reconstruction of Bantu ontology, psychology and ethics. When one goes through the book the common thread is to argue that there is nothing supernatural, mystical or animistic about the Bantu world – whereas it can neither be explained by our laws of physics. It does follow its own laws, which are explained in *Bantu Philosophy* (Tempels 1959 [1946]). What has made Europeans take the African world to be full of magic and mystery, is, according to Tempels, that the descriptions of the European anthropologists and missionaries of indigenous practices exoticized them, making them more mysterious than they were meant. One cannot say, of course, whether Tempels is 'right' here, whether he gives a true representation of the Bantu cosmology. What we can do is understand what he hoped to attain – a re-negotiation of reality that could repair as far as necessary, what had gone wrong in the racist othering of the Africans by the Europeans that came to colonize and use them. Tempels denies that Bantu ontology assigns 'force' as a separate predicate to a pregiven thing. All 'essences' are in or on the continuum of power or strength of life. Strength of life may be understood as interacting fluent energy, which makes all 'things' interdependent and in a continuing balancing act – that is the 'spirited' philosophy of Tempels/of the Bantu. In his words:

> Thus are, in the Bantu-philosophy, all the powers (essences)[10] of the entire world, not a multitude of unrelated, mutually independent powers, but all essences are interconnected ... from essence to essence. Nothing moves in this universe of powers without (potentially) touching all the rest. The world of powers is like a spider's web, of which no single thread can be made to vibrate, without making the entire web vibrate along with it.
>
> (Tempels 1946: p. 32, translation A. R.)

So what Tempels aims to do is to render the reality the Bantu live in/with, in a de-romanticized and de-occulticized manner.[11] This doesn't mean he gives a reductionist explanation of 'magic' in terms of modern natural science, but that he described a spirited way to be in the world, as in a world of essences, that hang together through vibrations of life-strengthening or life-diminishing actions, as an alternative way to describe what is experienced as real – as another form of natural science.

Notes

1 In this view Mosima follows Van Binsbergen (2003), especially chapter 15 '"Cultures Do Not Exist." Exploding Self-Evidences in the Investigation of Interculturality.'
2 As I could not find this edition in libraries or online, I make use of the Belgian edition of 1946.
3 Among those pillars are ideas such as the different substance of soul and body – classically used to undergird a Western Christian notion of salvation as eternal life in heaven after death. It is right that Derrida speaks, after Heidegger, of 'ontotheology' – as in classical Western theories of being and essence theology is never far away. All the basic metaphysical categories are defined to accommodate a unique, almighty and all-knowing God, as well as so many other tenets of Christian orthodox belief.
4 In this move, we see then, significant similarity with James's pragmatist pluriverse. James was no believer, he was interested in religion. As an outsider he saw that true religion brings bliss, and its opposite despair. To have true religion in this sense, he observes that we need no single universe defined by a system of theological dogmas about unity, good and evil, etc. 'Anything larger will do,' and multiple systems of ideas may do.
5 It is clear, however, that the translations can explain at least some of the critical remarks that black philosophers have brought against the work, and more, it can explain that such remarks are often accompanied with a general positive valuation of the book, as in Mudimbe, 1988.
6 Not a professional philosopher, but a Tempels reader.
7 Even though these may have to be revised later, when I hope to complete the mentioned entire new translation of the work.
8 Philosophically this is a region with a specific type of continental philosophy, which, next to the common continental philosophers from the tradition, built on

Hegel, Husserl, Heidegger and certain Catholic thinkers of those days who interpreted these types of phenomenology in a Thomistic framework.

9 Actually this entire paragraph misses from the English version, which only speaks of Western metaphysics, not of that of the 'primitives.'

10 Tempels provides the synonym here himself, which indicates strongly that he wanted to make clear that 'strength,' 'power' or 'force' is really identical with 'essence' – emergent forms of being. If being is the 'stuff' of everything, 'essences' are the ways in which it becomes or emerges in particular forms.

11 The same approach is taken by Ellis and Ter Haar 2009.

Chapter 7

Decolonizing nature

The case of the mourning elephants

Decolonizing human–animal relations

In 2012, elephants from two separate herds walked about 12 hours to hold what seemed to be a vigil for their deceased rescuer, South African conservationist Lawrence Anthony. Their story was met with reactions varying from intrigue to disbelief, as standing ideas on non-human-animals forbid us to think they might outdo humans in their capacity to sense the death of a close one, even across species-boundaries – not to mention that they would intentionally perform a ritual of mourning. In this chapter I will bring the critique of modern philosophical views of human–animal relations to the case of those mourning elephants, and will explore how alternative ontologies may be connected to it that express alternative relations to nature from an African context in our time, thus moving from Tempels's days to the present. At a point in time where environmental concerns are gaining more and more civil and public attention in African countries, environmentalism as such is still far from decolonized. Whereas mainstream conservationism is secularist, human-centered and technology-focused, it runs the risk of reproducing the hegemonic colonial attitudes that created so many of the environmental problems that face humanity today. It is necessary, therefore, to take a step back from the solution-centeredness of mainstream environmental discourse and investigate our ontological frameworks to find out if alternative perceptions of nature (in this case more specifically of non-human-animals) can be included in such discourse. To this effect I will critically address conservationist environmentalism focused on non-human-animals, and propose, alternatively, a dialogical intercultural frame of thought that addresses the politics of epistemologies, thus allowing the negotiation of varying views of nature. The upshot of this approach will be to forego the idea that any 'system of thought' can ever capture the essence of things in a definitive manner, while avoiding a relativistic position, and maintaining that the phenomena we perceive are real.

The story of the mourning elephants holds several important elements for our purpose. Their analysis and discussion may add to a decolonization of human–animal relations, which should be the foundation for a truly dialogical environmental approach. Decolonization will thus be extended from the sphere of relations between humans to relations of humans to non-human-animals (cf. Plumwood 2003; Roothaan 2017). The elements of the story under scrutiny consecutively are: (1) the elephants' behavior has to be understood in the historical context of troubled human-elephant encounters, as well as land dispossession in (neo-)colonial contexts; (2) their 'family relationship' to the person who granted them asylum in his private nature reserve invites to transcend the 'colonial' othering of non-human-animals; (3) the elephants' potential to sense the dying of a 'relative' asks us to acknowledge distant 'feeling' perception, which is acknowledged in traditional, 'shamanistic' or spirit ontologies. This chapter hopes to show that all three elements lead to understanding and accepting human perception and agency to be continuous with that of non-human-animals, thus contesting the idea of the human–animal divide discussed in Chapter 5. It will also be made clear that we can only arrive at such a view in a philosophically convincing manner by acknowledging the need to negotiate our epistemologies in the political realm.

Wildlife conservation aims at international, national as well as local policy levels, which have long stopped to be just an interest from concerned Westerners trying to address the side effects of global trade and industrialization in neo-colonial contexts. Even though a certain evangelizing and patronizing approach, smacking of the ideology of civilization that went hand in hand with colonialist projects, is not absent from the many reports of nongovernmental organizations (NGO)'s and intergovernmental organizations working for the preservation of the natural richness of the earth – the research as well as the preservation work itself is just as well initiated by politicians, entrepreneurs and academics from formerly colonized countries. This means that the 'decolonization' of which I speak here is not meant to make a plea for transferring initiatives and programs from the hands of former colonizers to the formerly colonized – that is already taking place. The point I will make here addresses the fact that traditional, spiritual, shamanistic[1] ways to relate to nature, with which conservationist initiatives may have to deal, especially when working in more traditional rural areas, have never entered environmentalist discourse as equal epistemic options. Before they can be considered thus, the dominant and dominating secularist worldview would have to be opened up to discuss the politics of epistemologies at work in the environmental discourse.

With politics of epistemologies, as explained before, I indicate the issue that certain descriptions of the conditions of true and valid knowledge dominate others by means of power systems regulating human investigation. While making a plea for an open, democratic discourse on knowledge within the limits of 'valid knowledge' hegemonic knowledge systems have always excluded criticisms of the conditions of validity of that same democratic forum. As a consequence, we may see elements of shamanistic ontologies enter conservationist reports, however only within the strict confounds of what counts as valid knowledge: that knowledge that recognizes the modernist categories of space and time and causality, and of mind and matter, and that can be transformed to technological prescriptions for conservation of certain species or landscapes that are negotiated with the economies of tourism, industrial agriculture, mining and the production of items for global and local markets.

We find an example in a report in which elephant protection plays a role. In a recent article in the *International Journal of Natural Resource Ecology and Management* Abugiche *et al.* describe that where traditional taboos to hunt and eat elephants are still remembered and partially adhered to (2017: pp. 64–5), this may be played upon to insure future protection of the animal. They describe this as follows:

> There is need for new and holistic wildlife conservation policies that will blend traditional systems of regulation, myths, rituals, and perception with existing wildlife legislation in the country to enhance conservation...
>
> (Abugiche *et al.* 2017: p. 66)

We see here, as well as in similar publications (cf. e.g., Hens 2006), that only the effects of traditional worldviews for conservationist practices are being considered, whereas the epistemic content of those worldviews as such, and their focus on a certain relationship with the elephants, is being ignored. This has several important consequences. Not only may it be questioned how taboos can uphold their power over people's actions when they are cut lose from the original epistemic frame from which they stem, but more importantly this approach leaves the idea of 'conservation' as such – an idea that functions within a (colonial) worldview of human control over the earth and its creatures – undiscussed.

Val Plumwood,[2] who has consistently discussed the politics of epistemologies, has offered, in a 2003 article, several conceptual instruments that may help to step out of what she calls colonial and centrist relationships. Centrism, in my own analysis, is the frame that treats a specific worldview as the center from which to understand the world – it floats on an implicit and ideally invisible power relationship – foregoing that alternative worldviews take themselves serious and can enter full negotiations with the center. This is how Euro-American hegemony worked and still works, be it through political, military, economic or cultural means. It makes those who adhere to non-modernist worldviews take an apologetic stance, or to present their knowledge systems in prefabricated categories of the center, such as 'mythology,' 'magic' or 'traditional.' A deconstructive approach can help us where we can hardly avoid such categories – to loosen their grip on our colonized minds. Such an approach is taken by Jacques Derrida in 'The Animal That Therefore I Am (More to Follow),' which was discussed in depth in Chapter 5 (Derrida 2002). Plumwood, however, doesn't take the road of deconstruction, but while accepting that philosophical discourse makes use of oppositional categories, tries to enrich the concepts used for the 'alternative' worldviews, thereby making them ready to hold to their own center from where they understand the world.

Thus she speaks, instead of 'non-human' nature of 'more-than-human' nature (in a bid to decenter the human) (Plumwood 2003: p. 52). She proposes that we should resist the 'backgrounding' of humanist centrism, and 'foreground' the more-than-human instead – which then leads to viewing the human being as just one of the different agents peopling the earth (Plumwood 2003: p. 61). In the context of Australian nature she insists that we refrain from speaking of it in terms of 'wilderness,' as this ignores the impact of indigenous peoples (their agency) on their environment (Plumwood 2003: p. 62). Here she touches on the interrelatedness of colonializing relations between 'centrists' and decentered 'indigenous' peoples on the one hand and the same relations between human and non-human (more-than-human) 'earth others.'[3] Discussing postcolonial and deconstructive approaches I showed earlier that where a Euro-American hegemonic outlook treats only certain humans (modern, white, especially male ones) as 'really' human, dehumanization of non-Western peoples reinforces the diminishing of their ontologies, especially where these recognize the personhood of non-human-animals. Plumwood shows how Eurocentrism also denies positive qualities to non-European landscapes. In her example it is the Australian landscape that is viewed '… as a deficient, empty land, a mere absence of the positive qualities of the homeland' (Plumwood 2003: p. 65). In fact, such emptying out concerns non-Western humans, non-human-animals and even non-animal 'others' such as rocks, plants and rivers:

> In the colonizing framework, the Other is not a positively-other-than entity in its own right, but an absence of the self, home or centre, something of no value or beauty of its own except to the extent that it can be brought to reflect, or bear the likeness of, home as standard.
>
> (Plumwood 2003: p. 65)

Although her account is of the Australian case, where the indigenous Aboriginal people have maintained that their experience of the world is one in which the land is central: the sacred and narrative subject from which human narrative and action depend, it can be taken as a paradigm to study other places where Western centrism has with success suppressed and ignored views in which the place of human beings is considered to be co-dependent with those of other beings. Especially her claim to treat 'Earth others' as agents and narrative subjects in their own right could help to philosophically open up to a story as the one which is taken as our point of departure – that of the mourning elephants.

Troubled encounters

When we investigate the story of the elephant herds that came to pay tribute to their rescuer Lawrence Anthony, it is important to avoid to let it 'Youtube-ify' our response in a simplistic emotional manner. The story namely has many elements that evoke the centrisms that are at the root of our troubled relationships with animals such as elephants. We (who take part in modern Western culture) like to cry for dead persons who remind us of our own beloved mourned ones, we are moved by the type of the rescuing hero – especially when he is a white male, who with his rational and responsible foresight, as well as his high morals, counters the cruel and chaotic effects of unbridled growth of human occupation of 'wild' nature in African societies, in order to preserve a beautiful animal that otherwise would perish. As sympathetic a person as Lawrence Anthony may have been (and we have to admire his openness to listen and speak to the elephants that were pushed out of their natural lands), when we focus on him as an elephant whisperer, or as a savior of natural wildlife, we tend to background the complexities of the settler society in which the animals in this case were pushed to aggression.

Anthony himself has stressed in his 2009 book on his work with elephants that the conservationist efforts he developed in his wildlife park Tula-Tula (the former hunting grounds of Zulu king Shaka, as he notifies us) are in cooperation with 'local people,' whom he tried to get involved in wildlife preservation by giving them jobs. The term 'local' should make us already suspicious though. Here we find a description that already marks certain people (and, implicitly, their aims and ideals as well) as 'local' – over against the 'higher' national or transnational efforts of people like Anthony himself. Of course it was modern transnational economies that brought wildlife under threat in the first place, through their colonialist endeavors, as well as through the legal and political structures they left behind. The question regarding elephant (or any form of wildlife) conversation should therefore perhaps not be how to integrate locals in growing industries like eco-tourism and nature preservation, but rather how the effects of national and transnational economies that aim to create 'progress' by furthering material wealth can be curbed as such. The elephant herd that was saved from being shot by Anthony, would not have been in their situation in the first place, had not a certain part of the human race at one point in history declared that it possessed certain pieces of land and could control all living beings (including 'local' human beings) that were on it. After that event wild animals have to be 'protected' by fencing off pieces of land, also with 'local' people on them, that are then artificially singled out from the 'normal' use and abuse of the earth 'outside' of the wildlife park.

When one zooms in on what lies behind so many animal conservation stories, like Hector Magome and James Murombedzi have done in their work on the political and legal issues at stake in the management ownership of national parks, the complexities on the ground come into view. Discussing several cases from countries in southern Africa such as Zimbabwe, Namibia and especially South Africa, they analyze the troubled relations between governments, private owners, local communities that form the net in which all together, the original human inhabitants, the newcomers, the non-human-animals and non-animal others are caught. When one looks at the very concrete historical circumstances that formed society in South Africa, one sees elements that prefigure the complications surrounding wildlife and nature reserves.

> [...] land dispossession in South Africa was based upon apartheid policy, a racially based separate development strategy that was designed by government to advance and benefit the interests of its minority white citizens at the expense of its majority black people. Although colonial influence in South Africa dates back to 1652, when the first European settlers arrived, the land conquest was institutionalized when the apartheid government passed the Natives' Land Acts of 1913 and 1936, which restricted land ownership by black people to just 13 per cent of the country's total land area. The land set aside for black people consisted of fragments scattered in selected areas of the country, first called 'native reserves' and later 'homelands.' This land was, with few exceptions, infertile and thus agriculturally unproductive. This situation forced many black males into the migrant labour system of the gold mines.
>
> (Magome and Murombedzi 2003: p. 109)

While land rights were differentiated as private (mostly white owners), national (the state, that under apartheid was organized toward benefiting the whites) or communal (good enough for local people), along an axial line of a center and its peripheries therefore – non-human others were kept out of the balance altogether. Until many species almost disappeared and people with power started to realize they had to do something about it.

In his article on the 'devolution' of wildlife management, Murombedzi provides further analysis of how postcolonial political, legal and organizational structures are no ideal frames to negotiate the needs of impoverished and often culturally and geographically uprooted 'indigenous' people, versus those descendants of the colonialists whose rights are often still served best by those structures. In South Africa e.g., policy makers have to deal with the dual system of land ownership:

> Southern Africa today, and especially Zimbabwe, South Africa and Namibia, is characterized by a distinctive dual land-tenure system, with

individual freehold tenure for a 'modern,' mostly white, farming sector and 'communal tenure' for the 'traditional,' exclusively black, farming sector.

(Murombedzi 2003: p. 138)

Present day wildlife management has to try to involve the rights and needs of local peoples whose traditional legal systems often aren't even officially recognized, which creates, next to population growth, the growing predominance of factors such as Chinese consumerism in African markets and an increasing desire of wealthy tourists from all over the world to see 'pure' 'African' nature. All factors together create complicated issues, in which wildlife itself is unwillingly entangled. It is within such contexts that the story of Anthony's elephants, who were threatened to be killed off because they were trying to escape from their original wildpark continuously and saved by him from that fate, should be understood.

Mourning elephants as moral agents

Environmental thinkers such as Plumwood and Harvey have shown us that you don't have to belong to an 'indigenous' people to be open to non-human others that speak and act, that tell about things, that mourn and perhaps even worship. In a world and time in which modernism has reached all corners of the earth, albeit to different levels, we are all subject to humanist centering and estrangement from nature, as present in our dependency of globalizing streams of consumer goods like clothes, processed foods and modern medicines. It is a complicated issue in general to get to really acknowledge non-human-animals as our likes, as agents and narrative subjects. Besides the growing material interdependency of peoples from all cultures, the cultural and religious missionizing projects that went along with colonialism have also led to curvy roads for those who want to recuperate alternative approaches to animals. An example is to be seen in the life of the former Congolese Catholic missionary in Mongolia Gaby Bamana, who, after long years of trying to convert Mongolians, came to the conclusion that he couldn't be successful in leading them away from their shamanistic ways – he couldn't do what Europeans had done to his own ancestors. This experience made him turn to a new career as a researcher in anthropology, trying to understand the spirit ontology of the Mongolian herders, that draws quite different lines distinguishing 'humans' and 'animals' than usual in the Western view:

> Analysis of research conversations I conducted and observations I made between 2010 and 2011 suggests that, in spite of the difference in species, herders considered dogs to be kin to humans (*neg yas*) because dogs are believed to share the same ontological nature as humans (*neg töröl*).

Thus, dog and human spirits are connected (spiritual analogy), and one practical implication of such connection is the social relationship of solidarity in everyday life.

(Bamana 2014: p. 2)

Bamana investigates why dogs have a special place in Mongolian herder culture. In contrast to horses, for instance, who are also favored domestic animals, only dogs have personal names. He finds that in a special way dogs are thought to be relatives of humans, and this not in an abstract metaphorical manner, but, it is believed, because they share their ontological substance, their spiritual essence, with humans. Dogs and humans, according to the herders, can be reincarnated into each other and share a mythological/spiritual descent. In the work of Kohn, we equally found an attempt at further philosophical explanation of interspecies relations in shamanistic cultures. Here too, we find human beings having a special relationship with dogs, who hunt with and for them (Kohn 2013: p. 131 and further). The people analyze the dogs' dreams as foretelling knowledge, like they do their own – and in doing so, they develop ways of communication that are open to different ways of giving meaning.

Elephants, one might say, are a completely different matter from dogs – canines have lived together with humans for tens of thousands of years, while elephants up to this day are so-called 'wild' animals. Wild here meaning not that they don't have socially regulated behavior among themselves, but that there is no standing social relation between their own and human societies. When we meet an elephant we will have to first negotiate how we will communicate with each other so to speak, whereas with domestic and farm animals there are already inherited patterns of communication in place.[4] In the story of Anthony's first encounter with the frustrated and angry elephants it is precisely his talent to do this which made him succeed in his effort to move the herd to his land where they could live without coming into further conflict with humans. In his book *The Elephant Whisperer* (Anthony and Spence 2009) it is described how Anthony, trying to convince the angry elephants to abstain from destructing the fence, speaks to the matriarch of the group, in English, making a guess that she will understand the tone of his voice, or that somehow his intent will come across. And it seems it did, because the elephants calmed down after he told them that their only chance at life and safety (their original matriarch had been recently shot) was to cooperate with him.

Of course many people are amazed at his courage to stand before an angry animal that could easily kill him, and even try to speak to it. We should not stop at such amazement however, but explain what it was that made his attempt at interspecies communication (to speak trans-species pidgin, as Kohn calls it – Kohn 2013) possible. One of the possibility conditions of such a communication should be the understanding, the belief or trust, that the elephant is not totally alien to me. That there is some kind of kinship, as we may think of it in evolutionary terms, that we share a biological ancestor and are 'built,' and wired, in similar ways – or in terms of a spiritual common ancestry as in the case of the Mongolian dog-herder relationship. Another is, however, the understanding that elephants have agency, and therefore, intentionality: that they can show deliberate behavior, and are not just driven by 'animal instincts' as philosophers and scientists alike used to ascribe to non-human-animals.

Nowadays science is also opening up to ascribing intelligence, intentionality, deliberation and agency to a growing range of species. In science, however, one can only maintain such ideas after hypotheses (that spring from theoretical renewals like the ones made by Plumwood and the likes) have proven to be true according to observed behaviors. Empirical evidence is the criterion of scientific truths. That is why modern cultures that take scientific results as their ontological measuring stick, have such a hard time to acknowledge and take seriously the knowledge of shamanistic cultures. Even though a shaman or sage from such a culture can say that he *knows* that humans and dogs, or elephants for that matter, are related spiritually (because he *saw* it in his trance vision) such experience is not considered to be empirical evidence. Empirical evidence is namely restricted to controlled and repeatable observations.[5] As Rose points out rightly, it is the epistemological questions concerning a shamanistic approach that are presently the hardest to answer – which made Plumwood choose for a future-oriented, ethical approach of non-human others, with whom we share being-of-the-earth, being earthlings:

> Most of her argument was laid out extensively in *Environmental Culture*. Here she put forward an interspecies ethic of recognition which depended on a particular stance toward the nonhuman world. That is, she was not making a set of truth claims about the world, but rather was asking what kind of stance a human can take that will open her to a responsive engagement in relation to nonhuman others.
>
> (Rose 2013: p. 97)

Although starting out from a concern about culture and identity rather than from ecological concerns, Nigerian philosopher Ekwealo also limits his revaluation of traditional approaches to nature to ethics. He calls his approach ecocentric and holistic, and propagates a new, decolonized environmentalism should be built on a new, Africanized, ontology:

> Consequently, a correction of all environmental and human associations believed to be progressive would start on a philosophical level in which there would be an exercise in deconstruction of earlier metaphysical thinking and re-construction based on a new ontological foundation. It

is only within this background that all ideas on conservation, sustainability, restoration and issues of peace, harmony and development would be possible.

(Ekwealo 2017: p. 15)

Similarly, Michael Eze makes a plea for a new ethical approach to nature, and calls his version eco-humanism. Like Ekwealo, he argues for an African approach to nature, which takes all being as being enlivened by vital force, a concept they both take over from Placide Tempels's *Bantu Philosophy* (1959 [1946]). As Ekwealo puts it, '… Africans believe that "force" or "spirit" is all pervading energy in the universe irrespective of the form or nature of its manifestation' (Ekwealo 2017: p. 55). I will not go into the question of whether Tempels gave a valid rendering of all-African metaphysical viewpoints, but will turn to the issue that all these proposals for a new, Africanized, environmental ethics, pass by the discussion of its foundational ontology from a critical epistemological viewpoint, including the political question as to who decides what is valid knowledge. Therefore this way out of humanist centrism lacks in a most important respect: that of challenging the politics of epistemologies that sustain the colonialist frames that still dominate how we view knowledge, and therewith, reality. While *animists*, such as Plumwood and others call themselves, or *holists*, as Ekwealo and Eze call themselves, only move in the realm of ethics – both types of approach leave the politics of epistemology to posterity. Revaluing the earth and its others, decentering us humans and re-centering the earth itself, they still leave politics aside, thus ignoring the struggle for 'truth,' a struggle which relates directly to political struggles as they are fought everywhere, be it against the Dakota pipeline in the US or the Bela Monte dam in Brazil, against the damage to original forest, done by large scale logging and farming from Asia to Central Africa. All these struggles are still related to the one about the truth.

Distant feeling: the reality of the spiritual

Let us return here once more to the question concerning my choice of the term 'shamanistic' over animist, for what I consider to be real alternatives to the modernist interpretation of reality. As Rose contends, the term animism originated with nineteenth-century anthropologists, who aimed to create dualist descriptions of 'civilized' and 'primitive' peoples. The primitives were the animists, who in believing that everything (not just sentient beings, but all natural phenomena) was enlivened by a spirit, failed to recognize the fundamental difference between mind and matter, and between humankind and everything else (Rose 2013: p. 96). Present day academic animists who self-identify as such in a oppositional manner rethink the view of animated reality in a positive manner. It is not so much its origins which made me move from describing my interest as animist to shamanistic – but the fact that it is an ontological indicator instead of an epistemological one. If we just recognize there to be varying ontologies (stretching thereby the claim present in the singular use of the term), going together with cultural differences, we might easily slip in cultural relativism, and its side effect of letting existing power relations between knowledge systems remain in place. The term shamanistic, however, points to knowledge (healing knowledge, knowledge of direction), and thus to epistemology. The shaman traditionally is the one who provides healing and direction for group members by entering upon a spirit journey, thus 'getting' knowledge in the spirit realm.[6]

In order to tackle the politics of epistemologies, we should investigate the truth conditions of the spirit journey. To do so we should develop a wider epistemological model than the Kantian one, which restricts validity to those insights that keep to the boundaries of space, time and causality (cf. Roothaan 2012: p. 120). Shamanistic experience, now, does not do so – as the trance traveler crosses space without reckoning with time, and vice versa. Also causality does not seem to play its normal role, as the healing procedures of shamanist practice involve the possibility to put things that happened in the past (even among the ancestors) right, or attract events from the future to the now, as in the case of rainmaking or the choice of a spiritual leader who is still an infant. The elephants in the case we are discussing now seem to possess shamanic abilities – sensing across a distance, and possibly (although the descriptions from several media are not clear about this) even before the fact, the death of their 'friend,' as they are supposed to have arrived at his house the day after he passed, having walked a long distance.[7] Such epistemic abilities have been recognized in certain African epistemological systems – like in that of the San, as described in the collection of myths and stories about animals in Africa by Shelagh Ranger. Citing Peter Garlake, she writes how animals are thought to have shared humanness with the human race, before God gave the different species their different roles and behaviors. Therefore

> Animals retain elements of their human past and nature; they conceive of themselves as human, are interested and involved in human affairs, will interfere in, help and hinder them. Animal behaviour is ... rational, purposive, directed by values and customs and institutions. Animals have language. Some practise sorcery. Their knowledge transcends that of humans in some areas ...
>
> (Ranger 2007: p. 80)

This description has also something very specific to say about our case, for it continues – citing San people remarking that ' "animals know all things," "they know things that we don't," they know what is going to happen: "an animal is a thing which knows of our death" ' (Ranger 2007: p. 80). This would mean that not only elephants can have such knowledge, but all animals. Stories about the behavior of pets living with humans tend to make one think in the same direction. Like the story of the two dogs of a friend's husband (who was terminally ill) – the dogs would come into his room regularly, but on the day that he was to die, they posted themselves at his bed, and stayed there the entire day until he passed.

It is an important question how this kind of knowledge can be considered valid – we need wider conditions of possibility than the Kantian ones that restrict knowledge to that which can be known through the natural (empirical) sciences. The epistemology to measure knowledge such as distant feeling and sensing death should take into account the wider reality of life: that of being in relation (as Plumwood also stresses) and that of realizing things – an epistemology which I have named pragmatic-interactive – it measures knowledge for its potential for action-in-relation, or its '... furthering (in more or less successful ways) life as it is shared, and at the same time individually enjoyed' (Roothaan 2012: p. 128). Such an epistemology concurs with ideas of Plumwood *cum suis* that knowledge should be practice- (future-) oriented and should center relations instead of a certain species and its interest, but it goes beyond claiming this to be a right (moral) *standpoint*. It claims also to demarcate what is true and what is false. The consequence of such a move is of course that many aspects of the modernist human-centered view of nature should be considered false, not furthering life as it is shared, nor as it is individually enjoyed, as they lead to extinction of our co-animals, to barren landscapes and to continuous sadness and fear in all those who do not belong to the centered group. This is the central warning in the stories of Yanomami shaman Davi Kopenawa:

> This is why we must refuse to give up our forest. We do not want it to become an arid land broken by muddy backwaters. Its value is far too high for it to be bought by anyone. *Omama* told our ancestors to live there by eating its fruits and his game, drinking the water of its rivers. He never told them to barter the forest for merchandise and money!
>
> (Kopenawa and Albert 2013: p. 281)

Sadness, anger about the impossibility to survive freely, according to tradition, was also happening to the elephant herd that was rescued by Anthony and his team. The turning point of the story how he rescued those elephants that broke loose continuously, and were about to be shot for it, lies in the efforts he made to talk to Nana, the matriarch of the herd, and to convince her that she should not have fear on his land (Anthony and Spence 2009: pp. 64–79). Thus, the story is about more than preserving a group of nine elephants. It counters the view that only allows elephants to occupy space as a beautiful asset to nature, or a superfluous hindrance to human society – in both cases ignoring their own agency and potential for knowing about their oppressed situation.

What, now, has been the result of this analysis, that aimed to critically address Western-dominated philosophical views of human–animal relations, and explore alternative ontologies for their potential to negotiate the environment from an African context. The story of the mourning elephants offered a good opportunity for this, as it finds itself at a crossroads of issues present in environmentalism. These issues are

a   the specific political, legal and organizational structures in previously colonized societies that form the frame of troubled human-elephant encounters, as it was shown referring to the works of Magome and Murombedzi.
b   the questions concerning human–animal relations as such: can decentering the human animal help to understand the elephants' behavior over toward Anthony as toward a relative?
c   the issue of the conditions of possibility (the politics of epistemologies) of 'shamanist' knowledge, and the kind of being aware of events beyond space and time as the vigil of the elephants seemed to imply.

In showing the intertwinement of empirical (historical, legal, political, etc.), ethical and epistemological questions, it became clear why ethical approaches to real world issues should always also include awareness of the political backgrounds of these issues. In our case these political backgrounds imply the politics of things as land rights and democratic representation, but behind that these imply the politics of what may count as knowledge. An epistemological approach that adopts conditions of possibility based on interaction and life-enhancement, instead of the Kantian one that bases itself on causality, space and time, might support 'decentered' worldviews, such as the one of the San that entailed the knowledge that 'an animal is a thing that knows about our death.' Recognizing animals (to start with, the elephants) to be capable of such knowledge, and of the intentional actions of compassion and giving back that are shown in their vigil for a human being they trusted, is then not just a nice ethical fringe on the dominant systems of power that go on to threaten life and well-being on this planet – but a critical act of resistance over against those systems. It would imply that we listen more to what animals, being obviously wise and caring beings, have to say to us. As a footnote to the above, we should recognize its implications – that not only non-human-animals, like elephants, are capable of spiritual knowledge that defies space, time and causality, but that we, being animals also, have similar abilities. It seems however, that for the most of us, this kind of knowledge is harder to access than it is for more-than-human-animals like the elephants, as all cultures have developed varying strategies for inducing trance and/or meditative states in which such knowledge first can be accessed.

Notes

1 A concept which can be used transculturally, cf. Roothaan: 2015, 141–2.
2 Crucial to her later work in environmental ethics has been her experience, in 1985, of being attacked by a crocodile in the Australian wetlands where she was canoeing. The rare combination of someone surviving such an attack, and that person being a philosopher, has left us with completely new anthropological insights on what it means that human beings are ecologically meant to be prey, just like other hunted animals. cf. Plumwood: 1995.
3 It should be noted, however, that the 'humans' here are the representatives of modernist cultures.
4 It is important to note that also a common pair of opposites such as wild versus domesticated hides a power relationship. In the case of the domesticated they are understood to work for free for humans, or give them pleasure. In the case of the wild, the absence of any institutional or 'political' relationship is again interpreted in the favor of the humans – as they see themselves as the ones responsible for the kind of relations that will be established.
5 In an attempt to open up modern science and philosophy to spiritual knowledge, William James therefore spoke of the need for a 'radical empiricism' which would take seriously all kinds of experience (cf. Bordogna 2008).
6 Less accurately called a trance journey. Although the shaman induces his journey by entering a trance state, this is not the distinctive aspect of the phenomenon. It could be theoretically possible for certain individuals to make a spirit journey without the trance – to get spiritual knowledge so to speak directly, as is said of individuals who are extraordinarily wired to look into the spiritual realm (cf. Borg 1994, about Jesus as 'spirit person').
7 Now it becomes clear why I preferred the expression 'spiritual journey' to that of 'trance journey' for we do not know whether animals who show such abilities need to experience in a trance state. It might be possible that they, and human beings too, have this kind of knowledge while not leaving their everyday state of awareness.

Chapter 8

Spirited trees – negotiating secular, religious and traditionalist frameworks

Intersecting frameworks

In traditional cultures all over the world trees have been important to humans as 'natural symbols' of the central values of communal life, as sources of food and medicine, and as signs of spiritual realities. In this chapter I will focus on trees in African contexts, where their meaning and place in the landscape and in human settlements are often contested in clashes between economic developmental projects, religious conversion movements, secular conservationism and traditionalism. These clashes will show more in detail what is at stake in negotiating the environment. It will include bringing very different discourses – those of traditional spirituality, of organized religion and of secular conservationism – to enter a common discourse. I will first outline which different frameworks determine the different relations to nature in the West-African context – these are those of secularization, of monotheistic theology and of traditionalism. While conservationists may work together with Christians, Muslims or those practicing traditional religions, their basic presuppositions will be scientific and secular. All the same, Muslims and Christians have their own spiritual understanding of the relation of humans to nature, in the frame of a theology of creation, religious morality and the afterlife, which can conflict with secular as well as traditionalist outlooks. All parties mentioned will also try to negotiate their interests within the framework of the global economy, 'selling' sacred forests as places of touristic interest or of rare plant life, for instance, or using religious fervor to promote economic progress.

Once the intersections of the frameworks for understanding the meaning of trees are delineated, I will discuss two examples of cases of the cutting of trees which show conflicting meanings attributed to trees. We will see how the frameworks clash into each other and make a shared discourse on what is of common human interest a complex issue. To view trees as a locus of contested meanings, it is important to understand them as bearers of symbolism and signification. A paradigmatic work of such nature has been done by geographer and Islamologist Eric Ross, who described the meaning of trees in public and religious discourse in Senegal. In his study of the holy city of Touba, the capital of Mouride sufism (which represents tûbâ, the tree of paradise), he shows how neo-platonic understandings of (spiritual) reality play their role in sufi theology, and how an ontology can be seen to be at work in it (as it is in Christian traditions that value worship of holy places, moments and persons). This kind of understanding and worship I will call shamanistic, as it opens up the kind of knowledge that shamanic practices do. This shows us that shamanistic ontologies are at work not only in so-called 'traditional,' pre-Abrahamic spirituality, but in monotheism as well. Through this exercise, we will see that religious ascription of meaning to trees is not to be understood in a simple, one-dimensional manner, but that here, as in other frameworks, complex historico-cultural developments have shaped interconnected layers of understanding and practice. In the conclusion of this chapter, I will include African environmental philosophy in the discussion of the different contexts and discourses within which trees are understood. As an example, I will discuss in some more detail Eze's work on eco-humanism, that was already mentioned in Chapter 8. In the tradition of Tempels and the Kenyan philosopher of religion John S. Mbiti, he proposes to adopt a holistic view of nature understood as life force, which leads to viewing human beings as an element in the whole of nature. Partly repeating my argumentations from the previous chapter, I will show that as much as a holistic account helps to understand the spiritual meaning of trees, it suppresses the difficulties that arise from the clashing of frameworks. Without an analysis of the effects of colonialism and Euro-American empire on the politics of epistemology, we may oversee the need to first get to a situation where negotiating the environment can be done fruitfully. We need a multidimensional approach, that brings the varying positions and views to a shared space of discussion, after first having disentangled and understood their differences and conflicts.

When we study the subject of trees in an African context, several frameworks of action have to be taken into account. There is, first, the framework of modernization, with its separate streams of secularization, the growth of an autonomous economic sphere and the separate but interconnected move toward ever more technological development. These streams are not identical, even though they have been going hand in hand for the most part of European history since the seventeenth century. In the United States, for instance, for a long time one of the world's driving nations with respect to economic and technological developments, society and politics remained largely determined by the Christian religion. On the African continent, things have been even more complex, and writing historiographies without a heavy bias of the colonial outlook is only still in its beginnings. Even though secularization is a force in present developments, the processes of Christianization and Islamization are just as important. Economic and technological developments very often are also driven by these religious developments, as we will see. The potential of different religious groups to impact society, in its turn, depends on economic and political factualities, that in themselves are not religious. The rise of societal influence of the Sufi-Islamic Mouride brotherhood in Senegal, for instance, could only take place in the political voids left by colonialism in their destruction of traditional kingdoms and nobility, as well as in the frame of economic opportunities created by the colonial trading system and the new global trading networks that came after decolonization (cf. Barry 1998 [1988]; Ross 2006).

Movements which focus on spiritual purification in a monotheistic context, and/or conversion of people, may coincide with a modern secular outlook, that values technological and economic progress. In concert, these may cooperate in overruling traditional practices that deal with nature as spirited. Consequently, overruling traditional inhibitions to use non-human beings for human ends. An example can be seen in how the Mouride brotherhood in Senegal, valuing prayer and work as central values, has played a pivotal role in cutting down several forests (which were traditionally seen as sacred/spirited places) for purposes of peanut production. All the same, whereas the Islamic theology in principle robs trees from the protection by guarding spirits (because the might of God/Allah will always be greater than that of any spirit), it still attributes a spiritual meaning to nature in general, and to trees in particular, as the symbol of heavenly paradise that Allah holds for his believers (Ross 2006).

A framework that is also of importance for understanding the plural meanings of trees is the modern secular framework that drives the work of many international environmentalist and conservationist organizations working on the African continent. Such a framework will make us see religious movements, as well as traditional beliefs and practices, uniquely from the perspective of their potential effect on the environment, without valuing their normative viewpoints in their own right – even if they are recognized for pragmatic reasons. As Siebert writes in an article on conservation of sacred forests in Benin:

> The new conservation trend assumes that biodiversity can be conserved by preserving culture and religious norms. Conservationists also often understand so-called 'natural' sacred sites as traditional conservation areas. Translating sacred sites in such a way often assumes that religious norms are readily compatible with conservation goals.
>
> (Siebert 2008: p. 165)

The environmentalist/conservationist framework however, brings its own normative views, of sustainability within the goals of modernization and economic development – norms that may easily overrule reverence of communities to trees, fungi, insects, rodents, humans and others making up the forest, not to forget the spirits. Sustainability means that human societies should not use up resources, while only looking for short term profit, thus creating long term death and destruction. All the same, propagators of this framework never ask about the perspective expressed in the word 'environment.' Although striving for a sustainable environment for humans will probably have the effect that animals and trees and other living beings also have better chances than if we only strive for profit and domination of nature, the perspective of secularism does not apply an intrinsic meaning to nature, or to trees, for that matter – its only frame of reference is humanity. It is an anthropocentric frame. So the idea that non-human beings, like trees and animals, have their own desires and values, that they deal also with us, while living on this same earth, will never enter the viewpoint of such an approach.

Finally, there are also traditionalist movements, which, in an attempt to counter the negative influences of the colonial era, aim for a return to traditional ways to live in and respond to the natural world. An example is Benin, where Vodun religion has been recognized once more on a national level, after colonialism suppressed it. Traditionalist movements have for their aim to preserve (or, if necessary, to reconstruct or even *reinvent*) traditional knowledge and practices in the frame of a spirited ontology. So here we have a specific situation concerning what to call 'indigenous,' as, formally, the settlers and/or colonizers have left – rendering the indicator 'indigenous' strictly speaking superfluous. Indigenous peoples now can be defined more precisely to be colonized peoples, whose self-rule is denied and who have limited or no land rights. In the case of modern African nations, there is a situation of self-rule of majority peoples – although the now independent nations in themselves are divided among traditionally living, 'indigenous' peoples (e.g., the San in South Africa or the Baka in Cameroon) and those living different versions of modern hybridized cultures. Among these latter mentioned, we find groups who aim to re-introduce 'indigenous' elements. In the present situation where the forces of global economies and nation-based politics reign, traditionalism will have to negotiate its values with those of, most notably, international tourism and national culture (cf. Juhé-Beaulaton and Roussel 2003), and through these, also with conservationism. We see this happening where the traditionalists' view of certain forests as sacred, has now been taken up by the international conservation efforts of the UN by declaring those forests cultural heritage, naming them the International Congress and Convention Association (ICCA)'s: Indigenous People's and Community Conserved Territories and Areas. When ways of life are taken up in the discourse of protected 'culture,' they may run the risk of evolving from ontological based relations to nature to forms of folklore that can be accepted in the frame of modernist ontologies.

Religion and the symbolism of trees

A first example of trees' contested meaning is provided in the case of the cutting of the Mbegué forest (Khelkom in Senegal), described by Schoon-maker Freudenberger (1991), a massive deforestation inspired by economic as well as religious reasons. These reasons coincided as 'hard work' in (agri-cultural) labor, which is seen in Mouridism[1] as a gateway to salvation. In this case we see how organized religion can suspend cultural or spiritual inhibitions against felling old trees while it may simultaneously stress their symbolic religious meaning, understanding trees as markers of the divine reality that awaits believers 'at the other side.' Schoonmaker Freudenberger noted in a passionate tone how in 1991 Mouride faithful, following a call of their religious leadership on the radio, cleared a whole forest in just a few weeks:

> From village and city, the faithful flocked to Mbegué, bringing axes and saws to clearcut 173 square miles of one of the last remaining forests in Senegal's degraded heartland. In three weekends, they felled more than five million Sahelian trees and shrubs.... For his part, the Khalifa-General (supreme head) of the powerful Mouride Islamic brotherhood was well on the way to meeting his goal: 45,000 hectares of newly cleared and fertile land would soon be put into peanut production.
>
> (Schoonmaker Freudenberger 1991: p. 1)

In this example we see a conflict of meaning given to trees by the secular environmentalist author (who is focused on the warding off of desertifica-tion, and upholding a livable human habitat) and the religious Mourides, who negotiated economic opportunities and a theology of the good religious life, resulting in ending the life of the forest.[2]

Another example of how the meaning of trees is contested between the different frameworks of modernism, monotheistic religion and traditional relations to nature, comes from a story narrated to me by Michael Onyebuchi Eze on the tree called Uvuru that used to dominate the central square of his native village. Ukwu Uvuru would have been more than 1,000 years old according to tradition, and had served from times immemorial as the place for public assemblies, as a source of fruits for children sitting in its shade, and it was ascribed a status of sacredness. It was cut down in 2002 by young Christians, who had been motivated to do so by their pastor. The motivation given was that the tree not just symbolized, but actually embodied evil forces that would explain the social and economic problems of the locals. It supposedly held the people chained to the past, and would hinder progress – so chopping it down was expected to solve that. Although further details of the story are unknown to me, it is a clear example of what happens when Christianity claims its ground by opposing traditional (folk) beliefs and practices. The idea is that ancient sacred places (or the opposite – places of evil), be they a well, a magnificent tree or a crossing of roads, should either be Christianized or erased from the landscape. In European landscapes the marks of this process in past centuries can still be seen, where a Christian cross, or an altar for a saint, mark the erasure (or the exorcism) of pre-Christian places that mediated spirit realities – often under an ancient oak, at a well, an offering place or at a crossing. The traditional, shamanic knowledge and practices that often survive in folk beliefs and rituals are seen as a threat to belief by many Christian pastors. In the example of Ukwu Uvuru the loyalty of people toward anything traditional, even if there might not be an attitude involved that would contradict Christianity, is seen as a sign of lack of religious trust – a situation that is combated by removing the physical body of the tree representing that loyalty.

In religious studies in general a sociological and anthropological point of view is predominant, which makes them study the resurgence of traditional beliefs in the context of the problems of modernization (cf. Juhé-Beaulaton 2006: p. 8; Ter Haar 1992: p. 111) – leaving the politics of epistemology that excludes shamanistic and spirit knowledge from the realm of valid knowledge out of sight. Economic difficulties, or the alienation resulting from urbanization, are presented as explanations for a recurrent turn to the spiritual realm of traditional deities. Christian theological discussions of the matter sometimes harmonize with these views in describing traditional spiritual practices as disturbing order in society. We find an example of this approach in a recent thesis on the tensions between belief in deities and Christianity in contemporary Igbo culture. The author, Christopher Okwor, writes as a recommendation that

> The church should as a matter of necessity adopt a holistic approach to evangelization. Christianity must be ready to feed the deep spiritual and material quests of the Igbo in order to control them and be able to divert their attention from deities (italics are mine).... Shrines are valuable heritage of our past.... Burning of shrines should be treated as a very serious offence.... Shrines should be developed into tourist centers. They are homes for endangered species of plants and animals and some have beautiful caves and springs.
>
> (Okwor 2012: pp. 134–5)

The central problem with traditional practices of worship of deities seems to be that they escape the dominant morality and legality and are therefore 'beyond control.' Here we see the secularist framework, with its two main interests, tourism and sustainability, join hands with a Christianizing interest: the striving for a moral life according to biblical prescriptions for the good life. We see a similar effect in the work of Juhé-Beaulaton, a historian, who directs attention to issues concerning law and order surrounding sacred forests and shrines. Like Okwor, she sees a harmonization between conservationalist, traditionalist and the legal and economic interests of the state as a solution to preserve the 'beautiful' places of worship while pacifying the spiritual needs of the people (Juhé-Beaulaton 2006). In these approaches however, the inherent value of the spiritual meaning of these shrines, often part of small forests, and therefore of the trees that mark and localize them, is lost, thus ignoring the shamanistic ontologies at work in the original use of such forests.

More apprehension for the religious and spiritual understanding of the world is found in the work of Eric Ross, who studied the symbolic meanings and usage of trees in West Africa, especially in Senegal, in their own right.³ Schooled in geography and (Islamic) religious studies, knowledgeable in Arabic and African languages, Ross represents a rare type of scholar who succeeds in combining disciplines to get to a deeper understanding of knowledge excluded from the white canon. While in non-African contexts the central tree in an African community has become known as 'palaver tree' (from the Portuguese word for speech or discussion), Ross indicates that this name is too narrow:

> [it] designates what is in reality a number of different phenomena which make political, social or religious use of individualized trees … Rather than a single 'palaver tree' serving as locus of public debate, polities were marked by a number of different trees, of various species, which served a variety of public and collective functions, only one of which was the 'palaver' process.
>
> (Ross 2008: p. 136)

Through field research, Ross has mapped and described ancient trees in Senegal that often have survived royal palaces or villages of which they once were the center. Thus, he opened up a new branch of research, combining the location on Google maps of the individual trees, photographing them and describing the oral histories that are told by the people living there. The interdisciplinarity needed for such research is important to bring localized, embodied knowledge to a wider sphere and make it relevant for intercultural discussions on ways to relate to nature. He distinguishes, with his informants, between the public, social, political and religious functions of trees. The ancient, hollow, baobabs are known to have been used as shrines, altars and tombs for the *griots* (court singers, like the medieval European 'jesters' or 'troubadours'). Trees are also mentioned as markers of historical battles, or ancient boundaries (Ross 2008: pp. 136–7). In their function of embodying political legitimacy they can be compared with obelisks, triumphal arches and other symbolic structures in Europe. In addition trees could be the places where justice was rendered. In these public functions they were and are seen as 'places of power' and places of memory (Ross 2008: p. 139 and p. 144) Trees thus function as

... markers, as memorials and as monuments. These functions are spatial in that they contribute meaning to the landscape, but they are also social and political, in that they 'fix' identities while also articulating a spiritual worldview.

(Ross 2008: p. 146)

Trees as archetype

In his work on the sacred city of Touba (an elaborated and reworked version of his dissertation), the center of worship in Senegalese Mouride Islam, Ross explains the mystical and symbolic understanding of trees. Mouride mysticism, defying any easy opposition between modern and traditional, combines elements of neo-platonic sufi understanding of the world in relation to God as creator and provider of grace for human beings to modern valuing of labor as a means of progress. The city of Touba, in the inner land of Senegal, was founded in 1887 by Shaykh Ahmadou Bamba Mbacké (1853–1927), on the authority of a revelation from God about the locality of this sacred place. According to a local legend:

After a patient search his [of the Shaykh, A. R.] attention was drawn, one day, to a tree which stood out clearly from the others by its size, its importance and its peculiar location. The tree ... stood on a plateau on the spot where the dome of Touba Mosque now stands. Ahmadou Bamba prayed in the shade of the tree, and that is where he had his long awaited revelation.

(Ross 2006: p. 28)

The sacredness of Touba is experienced and reinforced in the yearly pilgrimage of believers who come to offer prayers in its grandiose mosque. They long to be buried within the city when they die, as for them it symbolizes the closest connection to heavenly paradise. Touba, the name of the city, originally comes from the word tûbâ, which means the Tree of Paradise in Islamic tradition. It stands for the bliss of the heavenly state of the righteous after death.

As trees often play a critical role in foundation legends of places of human habitation, as well as of religious movements (think of Buddha who also received a revelation under a tree, and the symbolic 'tree' of the Christian cross), Ross explores the specifics of this role in the case of Touba within the context of sufi neo-platonic understanding. Here Touba is seen as

... a qutb, a 'pole' or axis mundi ... The concept of qutb is primarily an astronomical one. It designates the 'celestial pole,' a hypothetical spot in the sky around which the heavens revolve.

(Ross 2006: p. 18)

In neo-Platonism, the cosmos is understood as an unfolding of reality in layers of being, emanating from God, which are closer or farther removed from this, so to speak, ontological and spiritual gravity center. According to Ross,

> Sufism has invested this astronomical term with several related spiritual meanings. The term *qutb* is used to describe transcendence. It can be applied to any being, moment, event, or place which connects various layers of reality to each other.
>
> (Ross 2006: p. 18)

It can be a moment, which then should be remembered with reverence – like the moment of a divine revelation to an individual. It can also be a special, rare and beautiful place, such as a river, a grove, a high mountain or a city (cf. Jerusalem, which is understood as a place where the divine can be almost 'touched' by Jews, Christian and Muslims). It can also be, according to sufism, a person – in such a case the grave of this person can be dedicated as a place for pilgrimage. Touching the grave or some other act of dedication, is assumed to help the believer on his/her spiritual path. Ross further explains this in more philosophical language by citing Mircea Eliade:

> A sacred place constitutes a break in the homogeneity of space; this break is symbolized by an opening by which passage from one cosmic region to another is made possible (from heaven to earth and vice versa; from earth to the underworld); communication with heaven is expressed by one or another of certain images, all of which refer to the axis mundi, pillar ..., ladder (cf. Jacob's ladder), mountain, tree, vine, etc; around this cosmic axis lies the world (= our world), hence the axis is located 'in the middle,' at the 'navel of the earth'; it is the Center of the World.
>
> (Cited from Ross 2006: p. 20)

The *qutb*, the axis, can therefore also be a tree, and it often is. To make things more accessible for modern readers Ross introduces the term 'archetype,' coined by Carl-Gustav Jung in his psychology of the collective subconscious. He makes clear that one does not have to accept Jungian drive-psychology to use this concept, which just indicates what sufi theosophism means by fixed essences, the forms which 'hold,' or refer to, the divine presence of God. Although, as is clear from the above, designating trees to be an archetype of the relation of humanity to the spiritual realm is in no way restricted to an African context, trees play an important role in African spiritual symbolism (not excluding other spirited bodies, such as those of animals, rivers, etc.).

As Ross's research into the 'palaver' trees has made clear, the central tree of a village or royal courtyard often survives the mud-structures of the human habitats themselves. The trees thus serve as memorials of past human deeds and experiences. In the specific instance of Touba, the belief in the spiritual tree, symbolizing the entrance of paradise, is considered as the '...celestial register upon which the names and deeds of individuals are recorded' (Ross 2006: p. 31). The symbolism of the tree is even more elaborate, as the words of Cheikh Tidiane Sy make clear:

> In Islamic tradition, Touba also designates a tree of Paradise on whose leaves are inscribed each human's good and evil acts. Each leaf, as it falls, inexorably provokes the death of the individual whose acts have been recorded. The leaf is then preserved for the Day of Judgment.
>
> (Cited in Ross 2006: p. 32)

To inscribe themselves on this divine spiritual record, Mouride believers scratch their names and those of their loved ones on an actual Baobab tree in the city of Touba. In this religious practice there is thus no sharp distinction between a material act (taking out a knife and carving letters into the bark of a tree), and a spiritual act of faith (believing in or hoping for the saving grace of paradise, which is just another metaphor for the nearness of God). This 'mixing' of matter and mind, of belief and practice, shows how shamanistic ontologies are not just characteristic of pre-Abrahamic relations to nature, but also still at work in monotheistic religions.[4] Moving in the spaces of such 'hybrid' realities runs counter to the modern Christian (protestant) idea that only belief can save souls, for which reason much of traditional ritual behavior then should be considered irrational, and even banned. In this respect radical Protestantism reflects a similar spirit as secular modernism (although I would not want to make a historical claim as to one inspiring the other).[5] And since this secularism forms the actual (although not the necessary) framework of most modern philosophy, it has become hard in philosophy to understand archetypal relations of humans to trees as something more than irrationality.

The relation to ancient mysticism, which is so central in Mouride sufism, can be put in a wider framework, as Ross has done in an article from 1994, where he aims to understand present day Islam from an Afrocentric perspective. In Afrocentric research into the origins of African civilizations (originating in the work of Cheikh Anta Diop, and leading up to present day writers such as Sarwat Anis Al-Assiouty), the link with ancient Egyptian cosmology is increasingly researched, as well as its influencing role in Greek, Semitic and, finally, European and modern African science and religion. From a more philosophical perspective, this relation resonates in Mudimbe's work *The Invention of Africa* (1988), which, in a productive dialogue with the philosophy of the French philosopher Foucault, develops the understanding of African philosophy as gnosis, the neo-platonic word for higher wisdom, which can be sought by means of the secret training of the initiated. This concept of gnosis then, according to Mudimbe, is also a vehicle to transcend the cultural-anthropological concept of 'Africa' as a special place, or to transcend the localizing concept without leaving the place. Even to transcend the discipline and overcoming its colonialist preconditions by making it into '...a more credible anthropou-logos, that is a discourse on a human being' (Mudimbe 1988: p. 186).

Trees – contested meanings

After outlining the attitude toward the meanings of trees at the crossroads of conversion movements, traditionalism and modernization, we now have to take a closer look at the secularist point of view, before we can try to show how the different frameworks may be brought into dialogue with each other. The secular perspective which is central to most environmentalism as well as to the investigation undergirding it shows very clear in the reports of Schoonmaker Freudenberger cited above. In her account of the cutting of the Mbegué forest we see a shrill contrast with the tree-friendly mysticism of the Mouride brotherhood described by Ross. Schoonmaker Freudenberger shows in her work to have a more positive idea about the Fulbe pastoralists, who were displaced by the cutting of the forest, and who have been in conflict with agriculturalists in West-Africa several times. This choice is based, not on a more favorable view on one group's spiritual outlook or way of life, but on insights from environmental science and prospective expectations of which lifestyle in the Sahelian landscape might be more sustainable.

It is of importance here to underline the positive aim of secularist environmentalism – to preserve the earth for future generations of human beings, as a beautiful and healthy place to live, among a wealth of other species – plants and animals. From this aim for the future, environmentalism lends its legitimacy to critically question the behavior of human groups, peoples, but also governments and the corporate world. Its voice is an important one in the shared human efforts to live a good life, and its adherence to planning on the basis of scientific research provides an indispensable contribution to human relations to nature. All the same, one has to be critical toward its accounts too. First because all scientific results are provisionary, which often rest on researching a limited amount of factors, leaving others in the dark. Second because it fails to grasp the role symbolism and spiritual meanings may play in motivating human behavior. Third, because it doesn't question, in general, the neo-colonial conditions in which it does its work for 'the' environment (as it has been defined by modern Western science). If we put the second point to work with respect to the double-edged sword of Mouride theology – respecting individual trees as representative of the spiritual tree of paradise, while also stimulating agriculture as a good way of life for believers, even though it kills actual trees – we might be able to not just criticize the role of the Mouride leadership, as Schoonmaker Freudenberger does, for having too much power in a relatively weak state; but we might criticize it on a theological level itself.

How would such a theological criticism look like? It should direct itself at the level of understanding the relationship between the visible/material world and the religiously understood spiritual world (the paradise of God, the afterworld). In Senegalese Mouridism there might be a conflict at work between a shamanistic ontology which denies the dualism of spirit and matter, *and* a dualist ontology of heaven and earth as separate realms. This kind of conflict is not specific to this branch of Islam, but can be seen in monotheistic traditions in general, and even in modern secularist worldviews whenever they deal with the relationships of human beings to things. Whereas shamanistic ontologies recognize each (living) being as spirited, monotheist ontologies attribute a unique meaning to the human being as the servant of God, who plays a unique role in his plans with creation. The role the human being plays in monotheism is understood to take place in two realms simultaneously: s/he can live a moral life in the visible/material world, actually helping others or making the world a better place to live in, but the real meaning of these actions lies in the 'other' world, the world where eschatological issues play out. Issues that concern the 'salvation' of humanity as well as of all creation.

We find an example of this monotheist approach in the Mouride felling of the Mbegué forest, that lets the aims of spreading the faith and the striving of humans to deserve heavenly bliss in the afterlife prevail over the (spiritual-physical) connection of humanity with the trees concerned. All the same we saw the meaning of trees not to be absent in Mouride theology – as it is preserved symbolically in the reverence for certain exemplary individuals in specific holy places. Here we see how shamanistic ontologies may live in monotheistic religions, not even as a strange element,[6] but providing ways for religiosity to be experienced and acted out in human lives. Would shamanistic elements be more explicitly acknowledged for the ontological commitment they imply and the impact they may have on human behavior, they might be used to criticize and correct violent acts of religions against natural surroundings, thus harmonizing 'traditional' and 'modern' religiosity. Such an acknowledgment would also provide an alternative to the secularist discourse of environmentalism, while creating the terms for a dialogue on the religious interpretation of our environment.

A philosophical attempt to take the spiritual and the material well-being of human beings in relation to trees simultaneously into account, can be found in the article by Michael Onyebuchi Eze mentioned earlier, in which he offers an Africanist theory of environmental ethics. It presents a holistic viewpoint, that aligns with descriptions of African ontologies as given by Placide Tempels and John S. Mbiti. Inspired by their work, many African philosophers have presented a view of African religion, ontology and ethics in more generalized ways, in reaction to ethnographic work that provided descriptions of so many different African cultures as local and exotic, thus constructing an alternative to Western religion and ontology (cf. Ellis and Ter Haar 2009: p. 404). In Tempels's work the concept of life force has been highlighted. Tempels summarizes the 'Bantu' metaphysics as an ontology that encompasses all living beings and understands them as part of an interdependent energetic web. Whereas *Bantu Philosophy* (Tempels 1959 [1946]) claims to reconstruct reality as it is, it admits this reality to be accessible through different cultural frames in different ways. Following Tempels, Eze coins an understanding of ontology that is cultural as well as true – even if it may be mythological.

> It does not matter whether the exploits of Api or Uhere [river deities who were considered to protect the people, A. R.] during the Nigerian civil war is true or false. As a myth or legend however, what it does convey are certain philosophical truths ... of the people's cosmology and understanding of their environment. Their understanding and relationship to it is adapted through this socio-cultural mindset.
>
> (Eze 2017: p. 625)

Eze also quotes Ramose in this context:

> According to Ramose, this idea of community includes 'the greater environing wholeness in the sense of both the encompassing physical and metaphysical universe, together with the human universe in the sense of community.' This community is not just a collective of humans, it is a fluid habitation of interactive forces, beings, elements, animate and inanimate matters of the environment.
>
> (Eze 2017: p. 625)

For our purpose, to understand the meaning of trees, the most important conclusion of Eze's article is that 'environment' cannot be understood as the counterpart of an independent humanity. Nor can the aim of religion be taken to be opposed to a nature devoid of spirit. 'The sacredness of nature is because all elements have vital force' (Eze 2017: pp. 626–7). This leads then to the insight that 'The environment is not just inconsequential, it is part of life and constitutive of humanity' (Eze 2017: p. 627). In this view, characteristic of 'Ubuntu,' our relationship to trees should be caring and nurturing rather than economic, as the trees, together with all of nature, deserve our respect as they are among the life forms that sustain human life. An African philosophy of this kind is important for our goal, as it shows how the aims of environmentalism and a spirituality which encompasses respect and prudence in human–nature relations may be harmonized philosophically. All the same, Eze's approach passes over the necessary phase of disentangling the *practical* difficulties which arise out of the misunderstandings between the conflicting frameworks that dictate one's relation to nature. Only by articulating the frameworks, highlighting their loci of conflict and internal paradoxicality, can we come to articulate the conditions for a true negotiation of the environment, beyond the divides between secular sciences, religious theologies and shamanistic understandings and practices. In such articulation, namely, the analysis of power systems and the politics of epistemologies they engender can be tabled and faced.

Notes

1 Mouridism is an African branch of sufi-Islam, founded by the Shaykh Ahmadou Bamba in nineteenth-century Senegal. Bamba is revered as a spiritual educator and leader of a pacifist struggle against French occupation. Hard work (labor) and peace are important pillars of mouridist Islam. More below in the main text.
2 Below these conflicting views there is also a conflict between two peoples/cultures: the Wolof Mouride faithful and the Fulbe herders who traditionally lived in the forest, and who were displaced by the (legal) cutting of the trees. The Fulbe possibly also hold more traditionalist views of the spiritual meaning of trees, although they just as well consider themselves Muslims.
3 Note, this goes beyond describing meanings as 'emic' over against 'etic' – as this distinction between an insider and an outsider viewpoint still relativizes the ontology in question.
4 Even if full shamanistic ontologies do not focus so much on unity, as in the unity of God, but see all phenomena as potential manifestations of spirit, or as vehicles to reach the spiritual realm.
5 It can also be compared to modern forms of Islam, such as Salafism, which aims to erase shamanistic elements in Islam much like Protestantism did in Christianity.
6 Although modernization movements such as Salafism and Protestantism may view them as such – from their striving toward a purified religious tradition.

Chapter 9

Blurred, spirited and touched

From 'the study of man' to an anim(al)istic anthropology

White mythology decolonized

The roads that brought us to this final chapter have been winding, as they consisted of so many different argumentations for starting negotiations of the environment on several levels simultaneously. This aim has an urgency that cannot be deduced from theory, like when someone argues that science says the earth may become inhabitable because of changes that are measured in carbon-dioxide in the atmosphere, or rising temperatures in the global climate. Let it be clear, I do not intend to question these scientific arguments – I am not advocating denial of the results of scientific work on climate change. The point is that to really talk about the environment, like in a truly democratic palaver, all parties should be involved in its preparation. Also, scientific arguments that make us aware of environmental problems don't say anything about *why* we should keep the earth inhabitable for living creatures, and if so, which ones. Although science may not be as neutral as it claims, it is neutral in the sense that it defers questions of meaning and value. It doesn't ask *why* the earth presents itself to us like it does – at places brimming with life and available to intense vitalizing experiences, and at others still and barren – after it has been mined by humans or used for agricultural monocultures. Scientific discourse doesn't include earth, or non-human others in the palaver, presupposing they don't speak. When secular conservationism for instance makes a case that 'the great five' should be protected, it will not look into the interconnectedness with everything that would make survival of those animals possible – as they may need preserving the entire network of which they are a part. Instead we are often happy to preserve certain species in semi-domesticated ways, teaching them to live within wired areas, killing them when they try to break out of them or feeding them if their confines are not providing enough to eat. This is the case with the elephants discussed in Chapter 7.

To arrive at true negotiations we should start by looking honestly at the ways modern society makes us live. We should look into the ways colonials introduced legislation of land ownership while denying already existing indigenous legal systems. Into the ways modern liberal democracies even today denigrate indigenous education of children while stressing all children should get modern education in schools. We should look into the food chains that sustain us, and how we disrupt the food chains of others, spoiling soils, river systems, seas, mountains and the climate that interacts with them. We should investigate why we reject traditional methods of hunting, of growing vegetables, of birth control, religious attitudes to nature, ways to communicate with spirits, and with trees, for that matter. The necessity to negotiate the environment, and to do it now, is put upon humanity by the unacceptable violence in the relations between modern and indigenous peoples in our times, as well as in the attempts to reach out and communicate by some of their representatives, while in so many aspects colonial structures and institutions still rule their contact. More so, the spirit of modern cultures, their aim to institute a single rule of (modern, white) mankind over the entire earth and over all its life forms, is still very much alive today. It even works through conservationist attempts that aim to secure that same earth plus its inhabitants by controlling them from a modernist framework of scientific understanding.

Let me stress that the aim of this book is not to claim that peoples who uphold shamanistic cultures are in possession of the sole truth, whereas the moderns have it all wrong. Seeing the issue as a war of ideologies is getting it wrong. What is at issue is to create another conversation than the present one, a conversation characterized by openness rather than by the epistemological framework that only recognizes universalizable knowledge to be true. We need to include into the global intercultural dialogue those kinds of knowledge that are rooted in and from actual local situations. Indigenous peoples are *not* living in the past, as modernity wants us to think. The ideological claim of modernity, that there is a progressive history, and that the moderns are more developed than others because they struggle to remain at the forefront of that progress, is just that: *ideology*, by which I mean, the underpinning of a certain politics that treats 'us' better than 'them.' Even in their romantic moods, when they detest the artificiality of their modern lives, moderns localize indigenous peoples in the past, expressing how they, the romantics, long for 'ancient' wisdom and 'original' ways to live. We should denounce this kind of claim for what it is too: white mythology,[1] ideology that legitimizes the wrongs that are executed to sustain and promote the modern way of life. We are all contemporaries, and all bound up in transitory ways to find out how to live in this world. Moderns as well as indigenous peoples are in it together, and neither of them should claim to be in possession of 'the' truth.

Still, the evaluations made by indigenous people of the modern ways to be in this world are seldom heard in academic or public policy discourse. Whenever they are, however, they may sound something like the words of

Kopenawa, upon one of his visits to New York, during which he saw the poor areas of the city as well:

> These white people who created merchandise think they are clever and brave. Yet they are greedy and do not take care of those among them who have nothing.... They do not want to know anything about these needy people [...] and let them suffer alone ... and are satisfied to keep their distance and call them 'the poor.'
>
> (Kopenawa and Albert 2013: p. 349)

Of course his words are colored by his experiences in Brazil with the gold diggers who disrupted the life of his people. He denounces the white love of money, the white people's lack of social responsibility and their killing of indigenous peoples to get hold of the land they sit on, in order to cut up the earth, to destroy the forests and to spoil the environment, only to create more 'wealth.'[2]

From the other end, there are internal critiques of Western moral argumentation, such as the one by Margaret Urban Walker, who pointed out that we should move toward treating moral philosophy as situated practice, to avoid a false universalizing that modern philosophical ethics represents:

> Moral theorizing within the theoretical-juridical approach typically universalizes and homogenizes 'the' moral point of view or position of 'the' moral agent, and traffics in claims about 'our' concept of responsibility, sense of justice, intuitions, or obligations.
>
> (Urban Walker 2007: p. 18)

Such universalizing occupies violently the moral position of the oppressed – with the intent to make it hard for them to realize they are not considered to be part of the moral 'family' of their oppressors. The dominant modern ethical theories '... effectively erase the majority of human beings in depicting a moral persona or identity dominant in some form of social life as if it were the only one' (Urban Walker 2007: p. 18). We need refinements like those made by Urban Walker to be able to see the important difference of being among those for whom the system was created, so to speak, or among those who are considered of a lesser kind and are made complicit by pushing a false consciousness upon them.

In his *Wretched of the Earth* (2004 [1961]), Frantz Fanon addresses those who are captured in such an alienated consciousness – particularly those who have, after decolonization, entered the phase of living in neo-colonies that are still under an economic reign imposed by the West. Like Urban Walker, he makes it clear that a 'moral we' can be the trick of the oppressor to join him by believing in his deceitful stories of 'human progress':

For centuries Europe has brought the progress of other men to a halt and enslaved them for its own purposes and glory; for centuries it has stifled virtually the whole of humanity in the name of a so-called 'spiritual adventure.'

(Fanon 2004 [1961]: p. 235)

This Europe, which never stopped talking of man, which never stopped proclaiming its sole concern was man, we now know the price of suffering humanity has paid for every one of its spiritual victories.

(Fanon 2004 [1961]: p. 236)

Fanon calls upon his readers (and his intended readers are the oppressed) to stop being complicit in their own oppression by upholding their belief in the West's stories, and their liberating promises. Today these stories would be those of freedom and liberty, promoted as making anything possible – stories that include the promotion of education for girls, the support for democratic elections in neo-colonial areas or promoting free trade – while excluding those liberating policies that don't suit the West, such as land-reform, reparations or standing up to neo-colonial conservationists. In all, the ideal of living a life as a healthy, educated and 'free' consumer, is the norm. This Western ideal will be the norm for some time still. Not many have listened to the call Fanon wrote down in 1961, months before his death, that 'European achievements, European technology and European lifestyles must stop tempting us and leading us astray' (Fanon 2004 [1961]: p. 236). His call to 'endeavor to create a new man' has proven even harder to follow, it seems.

Since the times of institutional decolonization, critical philosophy in the West developed black consciousness, feminist theory and postmodernism. Their most important contribution to the dethronement of the ideology of the consumerist ideology is to have ended the talk of 'man' as a single and abstract category. If we do not radically acknowledge that human beings are *plural*, we cannot think clearly about how we as a network of human cultures want to move on. One thing that was also very clear to Fanon, and we will have to take it as one of our axioms here as well, was that moving forward should not be a matter of choice between development or traditionalism:

But what matters now is not a question of profitability, not a question of increased productivity, not a question of production rates. No, it is not a question of back to nature. It is the very basic question of not dragging man in directions which mutilate him, of not imposing on his brain tempos that rapidly obliterate and unhinge it.

(Fanon 2004 [1961]: p. 238)

If we compare his address to the inhabitants of the then newly 'freed' African states to the reflections of Kopenawa (representing a people that has yet to be recognized as having rights to land), it springs to the eye that the Yanomami shaman stresses life 'in the forest' as the life which has the right pace for leading a human life, and a life which makes clear thinking possible. It has to be noted once more, against hippie-romantic views of indigenous life, that that kind of life is not to be characterized as a turn 'back to nature,' even though it may include being closer to other natural beings than in modern city life. Indigenous societies are just as well that – societies, with culture, religion, legal traditions, medicinal knowledge, educational traditions, etc. This recognition is the foundation for any potential intercultural dialogue. To get to such a dialogue however, modernity and its defenders will have to be ready to stand trial before their victims – indigenous peoples, but also elephants, trees and spirits, including those who have been killed and come back to haunt. Without such a trial, dialogue will not make sense, as there would be no guarantee whatsoever that modern societies will give up those policies that destroy the ways of being in the world, and the peoples who defend them, that are called indigenous.

Blurred boundaries

Before the moderns can be allowed to negotiate with other cultures, they will have to concede first that their culture as such is in a war with indigenous ways of life, and the peoples who practice them. They will have to accept that genocidal actions are part of their own culture, as well as the systematic destruction of the natural environments of those it wants to transform to its own kind or let them die. They will have to be ready to reconsider their way of life, which can not be had without destruction and death. Finally this means they will have to be ready to give up their most cherished possession: their self-image as 'man,' the supposedly most worthy species of the earth. Whether because it supposes to have a special treaty with God, or a special faculty, reason, white humankind has implicitly treated itself as the best that ever happened. It has to be ready to give up this idea of himself as a unique kind that deserves to make the rest of all the earth's creatures into its serfs or to be used up and thrown away as garbage. It has to be ready to concede that its rationality is not unique, but that its boundaries are blurred, and that the image of 'man' is unclear. Being human overlaps with the being of many others, spirits, animals, even plants, fungi, bacteria, minerals and whatever elements we may be able to name or specify.

That we share much with all these others *was* clear to the Enlightenment philosophers, but they centered this one kind, mankind. Kant understood anthropology to include the study of all things in the world, and of the world itself – as cosmology, therefore, and ethology, mineralogy, etc. All would-be forms of anthropology, insofar as it is theoretical (we would now say empirical):

Such an anthropology, considered as *knowledge of the world* … is actually not yet *pragmatic* when it contains an extensive knowledge of *things* in the world, for example, animals, plants, and minerals from various lands and climates, but only when it contains knowledge of the human being as a *citizen of the world*.

(Kant 2006 [1798]: p. 4)[3]

Understanding the study of natural phenomena as comprised in the study of man, transforms all being into *environment* – into the background, the nurturing locus of humanity. This Kantian approach, called 'backgrounding' by Plumwood, is still taught in philosophy curricula as the newest Copernican revolution: where before the Enlightenment humanity was already seen as the special creature in Western philosophy, it was still considered to be understood within a cosmology dependent on a divine creator, and therefore as an element of this cosmology before being thought in its own right. Now, the tables were turned, as the entire world was understood to be *knowable* only by this rational being which man embodies – and that all that was not knowable in the right manner, although it was thematized by spirit seers, and natives of primitive lands, was rejected as being part of the world as we know it. Man, the one who knows, then becomes the center of everything, and knowing all there is to be known is knowing him. This view of things – even if the Enlightenment is tradition itself now – still rules the way the moderns relate to the world, and to nature. In our industries, our way of making our homes, our feeding habits, feasting practices and in our species' ambitions, such as conquering space, we still behave as if the rest is just that – a rest, disposable when not usable by modern mankind.

Next to this view of all empirical knowledge as knowledge of the human being, Kant distinguishes pragmatic knowledge, which has an entirely different character. The pragmatic anthropology Kant mentions here is the study of what man can *become* – it is a moral philosophy, based on a teleology of the essence of humanity. According to Kant we need to find out what is our deepest aim as human beings, to consequently design a system of education which can help us transform ourselves with an eye to that aim. For Kant this deepest aim was rationality, which he defined as *public* and *free* thinking. Thinking, being prepared to stand before the court of public reason, the reason shared by all rational creature, makes it free, in this view. It is considered to be free from loyalties to a tribe, a king, a family, to all the kind of clientele-like relations we may be entangled in. This definition of modernity as being free from local ties is considered to be one of the greatest accomplishments of Kant – although he took back with one hand what he had given with the other, when he decided that certain 'races' of human beings could not possess that rationality as well as Europeans could, laying the 'burden' to educate them with the white man. This modern anthropology, centered on the continuous self-transformation of man and the use of all who do not participate in it naturally, is at stake in the negotiations we are aiming at here. It makes up the treasure of white, modern ideology.

As I showed in Chapter 6, we can critique the modern approach to anthropology, in two ways – by decolonizing or by deconstructing the idea of humanity. The decolonizing critique founds itself on the reclaiming of humanity by the oppressed, by those who were robbed of it in the 'Copernican' move of the Enlightenment. The deconstructing move follows another path: it unhinges the argumentative connections that uphold the architecture of modern culture and thinking. Consequently it will also unhinge this humanity as being the well-defined category it is taken to be in modern thought. It will show that it is blurred, by questioning the possibility to express it in this singular word, 'man.' Deconstruction's effects also coincide with many of those brought about by 'feminist' thinking – and include the necessity to speak, from now on, of human beings in the plural, as wo/men – gendered, plural, situated, embodied embedded in networks of many kinds. Beings with blurred boundaries.

Dreamtime knowledge

The essential point for readers that need to decolonize themselves (whatever their origin) is to realize that Western philosophy and science are just one attempt of human beings to deal with life and death. And that Western culture, including its high level of technology, is also just one way of living life and doing things – it is not the end of history, not the highest culture ever, and the immense quantities of species members it has been able to sustain over the past two centuries is not a sign of its superior nature. All such characteristics are seductions to stop thinking, to avoid seeking truth. We will never be able to think clearly, or to see truth, unless we – even if we should try do so for the sake of experiment – succeed to oppose the seductive self-promotions of modernity. Only by opening up to the suffering created by this culture, our own suffering and the suffering of others, can the ideological façade crumble. Even if moderns succeed to do so, however, this doesn't mean one is getting 'pure' access to a 'purely other' way of being in the world. Bruce Albert, who, as a white anthropologist, tried to open up to the life story of his friend Davi Kopenawa, makes it clear in his explanatory afterword, how complicated it can be to get to a true conversation across the limiting conditions of the postcolonial situation in which we are all caught up:

> But something more fundamental than vigilance about epistemology was at stake in the intense 'ethnographic situation' in which I became involved.... What does it mean to be 'adopted' by one's hosts, when they see themselves increasingly subjugated by a threatening outside

world from which the ethnographer is essentially an emissary, no matter how laughable or inoffensive he may seem at first sight?

(Kopenawa and Albert 2013: p. 431)

In fact, Albert understands, the ethnographer is 'used' by those studied as an object of study as well, to better understand the world that is threatening theirs – and to use him/her, if possible, as a kind of ambassador in that outside world as well. In this process of defense against genocidal destruction of their way of life *and* their *actual lives*, their understanding – naturally – changes as well:

> What's at stake for his interlocutors is accepting a process of self-objectification through ethnographic observation in a way that allows them to claim cultural recognition and empowerment in the opaque and malignant world trying to subjugate them.
>
> (Kopenawa and Albert 2013: pp. 432–3)

When we, in our turn, try to listen to the voice that speaks from this book, the voice of Davi Kopenawa trying to understand the war he and his people are in, to understand his enemy and to defend his people – it springs to the eye that the moderns are mostly pitied for their lack of clear thought. This is not what they, encapsulated in their ideology of greatness, would expect. Their attitude is rather something like: ok, we created pollution as a side effect of industrialization, we will clean it up; or: yes, we have a lack of social care, and we may have lost the talent to live in egalitarian and caring societies, but our knowledge, our scientific research, our rational analyses, these we will always defend as being the highest achievement of mankind.

For a defender of modern epistemology it will be shocking that not only in their social and environmental behaviors, but even in their ways to acquire knowledge, the moderns are criticized by those who oppose modernism. Indigenous epistemologies, that take a connection to the spirits as the possibility condition for valid knowledge, cannot see how the rejection of spirited knowledge can be taken seriously. The connections with the spirits, namely, help the knower to distance him/herself from direct needs and interests, to see beyond the short term. This shows in one of the observations of Kopenawa:

> In their cities, one cannot learn things of the dream. People there do not know how to bring down the spirits of the forest and the animal ancestors' images. They only set their gaze on ... merchandise, television, and money.
>
> (Kopenawa and Albert 2013: p. 356)

To his view, the spirits are active in helping humans to achieve knowledge. Spirit-based knowledge, in all its different forms, seems always to have these

intrinsic aims: to *heal* and to *see direction*. Both are dependent on an understanding of reality that takes all creatures as interconnected, and this interconnectedness itself as the basic structure of reality. For those whose life
consists of growth and movements it is of utmost importance to see direction, and to live in a web of relations that is harmonious, not hostile. This is
given with the way our anim(al)istic body-soul natures are preconditioned.
The aspect of healing has its eye on the harmony of the web. Indigenous
knowledge in different forms claims that we get sick when the web is not as
supportive as we would hope – when it is hostile, eating away at this fragile
structure we call our own (body and psyche or soul). Shamanistic ontologies
stress that healing should never be directed only toward the individual who
is sick, but to its environment as well. The second aspect – to see direction,
is necessary because the world of relations in which we live and move is too
complex to oversee. The Western approach promises, in its claim to universality, the possibility to get a bird's eye view of things.[4] Shamanistic philosophies take such a promise to be utterly wrong. All knowledge is limited,
by nature, and therefore we always need 'direction.' Direction is not the same
as a blue-print or a clear telos, it is no hoped-for eschaton. It is a relative
thing, of which we ourselves, after receiving it, will have to find out how to
put it to practice and change our live or our world to follow it.

The healing and directional knowledge that takes the whole into account,
a whole which by necessity one can never oversee, needs some kind of 'superhuman' power. And here is where the spirits come in. They are not supernatural, but they are superhuman. They open our minds to give us insights
that Jung called 'subconscious,' but that might as well be named 'supercon-
scious' as they transcend the consciousness of daytime experience. In shamanistic epistemologies, the idea of 'dreamtime' (their name for this kind of
experience) is very important. Robert Wolff recounts of the Sng'oi in Malaysia who, each morning when they wake up in the common sleeping room,
share their dreams with each other to get direction:

> They did not think that they were sharing dreams as we think of
> dreams. The Sng'oi believe that the world we live in is a shadow world,
> and that the real world is behind it. At night, they believe, we visit the
> real world, and in the morning we share what we saw and learned there.
> The story that was created around the memories that four or five people
> brought back from the real world set the tone for the day.
>
> (Wolff 2001: p. 88)

Interestingly enough this view is exactly what Kant, in his *Dreams of a
Spirit-Seer* (2002 [1766]) sets out to oppose. He denounces dreams as a
legitimate source of knowledge, as Descartes and other modern European
philosophers had done before him. Their view goes back to the ancient
Greek mistrust of sense experience – for things we see or hear are always

changing and therefore unreliable. Also our senses may distort how things 'really' are. This is the more true for dreams. If sense experience delivers the shadows of objective existences to us, dreams provide the shadows of shadows. With all their strange changes of place and time, faces of people that may just change in the middle of a story, dreams fascinate and confuse the modernist mind.

For Kant dreams are purely individual, subjective, experiences. They stand for that which can not be shared or defended for the court of public reason. So the title of his work against Swedenborg – *Dreams of a Spirit-Seer. Elucidated through Dreams of Metaphysics* (see Kant 2002 [1766]) – says it all: like the dogmatic metaphysicians, who make up abstract concepts for realities nobody can check empirically, the spirit seer makes up his own realities, which he sells for spiritual truth. To back up this view, Kant refers to Aristotle, the paradigmatic no-nonsense philosopher, the one who defended early in European history the sound combination of reason and the senses:

> Aristotle says, somewhere: 'When we are awake, we have a common world, but when we dream, each has his own.' It seems to me that one should perhaps reverse the last propositions and could say: When different people each have their own world, then it is to be supposed that they are dreaming.
>
> (Kant 2006 [1798]: p. 29)

Within the class of dreamers, Kant distinguishes between the dreamers of reason (the metaphysical philosophers) and the dreamers of sense – who live by their fantasies. Apart from these two, again, stands the spirit seer. His visions must follow, according to Kant, from a physical rarity, through which someone gets convincing impressions of things that do not have objective reality. So the clairvoyant is not deceitful, but deceived by his brain. Consequently, this *waking dreamer* is considered to be mentally abnormal.

Analytically, Kant distinguishes night dreamers from day dreamers and dreamers of reason from dreamers of sense – an analytical approach we do not find in shamanistic cultures, which rather see a continuum between all the different dream-like revelations of knowledge. We can see this in the description of Kopenawa of his inability to sleep, when he stays in modern cities. The noise and spiritual chaos around him make him go into spontaneous trance states in which *he* sees the spirits of the whites:

> I slept in a ghost state and was often overcome by dizzy spells. Then, as in the other big cities I visited, I saw the *xapiri* of these lands of the ancient white people come down in my sleep.
>
> (Kopenawa and Albert 2013: p. 350)

Upon returning home he starts to drink the spirit brew once more, which makes him connect with the forest spirits, and re-find his emotional peace. There are many ways to 'dream' according to him – in sleep as well as when awake, and the best one is the one of induced trance of the shaman. To *describe* such ways to know has never been a problem for the moderns, as long as it is done in cultural anthropology, and when it is made clear that these are exotic ways of being in the world – remnants from earlier ways to be human. To understand this kind of knowledge as an expression of *culture* disarms it, avoids the question concerning its truth. To aim at intercultural dialogue, however, moderns should ask themselves whether they can understand its content, whether they can have similar knowledge, or, whether the parties can enter first into a dialogue about epistemology with each other.

Until James tried his hand at it, no serious efforts were made in modern times to *understand* dreamtime knowledge *philosophically*, that is: *as knowledge*, for that would mean one has to doubt the two dogmas of reason and the (non-deluded) senses. As Kant put it:

> For to wish to offer in a serious fashion interpretations of the figments of visionaries instantly arouses grave doubts, and the philosophy that allows itself to be caught in such bad company places itself under suspicion.
>
> (Kant 2006 [1798]: p. 35)

This book has tried to start and promote just such a philosophy. It tabled, first, the power relations that deny the epistemic claims made by shamanistic knowledge. Thus it challenges the universality of modern methods of knowing, and brings the politics of epistemology that is always at work into view. Second it tries to de-essentialize the idea of distinct cultures. Cultures are, as Mosima and Van Binsbergen argued, permeable – and they negotiate whenever they come into contact. They negotiate their ways to be in the world, to adapt to it and to change it to a state they see as good. Only if there is a (silent) war going on, in which one party is attacked to make way for the goals of the other,[5] negotiation has to be preceded by steps to make a more or less equal exchange first possible. Such steps are the hardest to arrive at in our present situation. They imply a generalized decolonization of Westernized minds, and the willingness of the moderns to negotiate what they consider their basic rights. Implementing these steps is not impossible, though, if the awareness will grow that modern ways to be in the world are essentially limited, and cannot be expanded endlessly. If the moderns start to listen to their internal critics as well as to the indigenous 'ambassadors,' they may open up to what haunts them, to their neglected spirits. The natural spirits that were forgotten, and the spirits of deceased victims of modern greed. Only if all these and more preliminary works will be done, intercultural dialogue can be taken up seriously. Intercultural philosophical dialogue is where we are headed, but it is the end of the process, not the start.

Humans, animals and all of us

Ontologies differ considerably, and they differ among those cultures we call modern as well as among those we call indigenous. I follow William James's solution to the problem of ontological pluralism, which entails that as long as we don't want to adopt a block universe (a metaphysical system that can only allow one way to be, and that therefore answers all questions concerning being), we should understand the world as a network of interconnected, while differing, ontologies. Intercultural studies confirm this to be the case, for we can understand different ways to be in the world and to relate to beings around us differently, while still discussing them in a shared discourse. We can understand the Mongolian herders' option to see dogs as related to humans, but horses not. We can understand Harvey's rendering of rocks expressing the history of the land in Aboriginal country. We can take serious how trees, or their spirits protect people, make space for their palavers and preserve the heritage of communities. Ontologies do not necessarily have to be defended for the court of a universal public reason to be counted in as options for human societies to express the reality they live with. We do not need to hold those who get knowledge from spirit entities, like Davi Kopenawa, Malidoma Somé and Robert Wolff to have deformed brain functions. Actually their works show that on the contrary they express insights into living in and with nature that may help to understand the human condition at a deeper level than the modern sciences do. We can accept that different ways to relate to nature overlap and that we need to find negotiated agreements to move on as interconnected communities of culturally differentiated peoples living together on the same earth.

The challenges of intercultural relations over the past centuries have been overwhelmingly on the side of indigenous peoples, who have struggled to survive, physically and culturally, over against the immense forces of industrialization, population pressure and the building of modern settler states on their lands – states that have been constructing layers of institutions (political, legal, cultural, economic) over their own societies, impairing those to function healthily. Now we live in a global postcolony, meaning that after the political process of decolonization, we find ourselves in ever so many economic, political and cultural constructs and processes of exchange that still inherit traits of colonialism. The process of repairing the damage done to the human network has only just begun, as it shows in hesitant moves like the return of looted art by former colonizing states, or respecting indigenous land rights or traditional legal systems in settler states like the US, Australia, southern American states and apartheid South Africa. The relations between new nations to their former colonizers, as well as those of old settler states to their indigenous populations, are still far from being decolonized. Most of the indigenous peoples do not have full rights over their own land, and their legal and spiritual systems are yet far from being fully recognized, on a state as well as on an international level. Their medicinal knowledge is often patented/appropriated by pharmaceutical companies. If now some representatives of the moderns, are getting ready to enter into dialogue with representatives of indigenous, colonized and oppressed peoples, a challenge arises for modern cultures – for such dialogues cannot be without consequences. They may have for a consequence to limit the rights of modern states, to restrict its claims to land, to allow for different models of legal systems, different forms of being family, including ways to distinguish genders, etc. Most of all, it will certainly have for a consequence that moderns need to limit their use of mineral energy, of water, electricity, of meat and of vegetal agricultural products for which original forests go down, their spirits are contested, and their animal life is ignored as having rights as well. A recognition of what it takes to truly dialogue, will have for a consequence that subjects such as energy, food, population growth, migration, have to be tabled in their interconnectedness with cultural, religious, spiritual and ontological differences, instead of as problems for the world as seen by the moderns, as is the dominant view at present.

This book will have fulfilled its aim if it has made clear that at least part of the answer to such challenges doesn't lie in more planning and more science-based policies, but in opening up to other ways to gain knowledge. The shamanistic way to find knowledge in the dreamtime lets people understand the world in ways that, despite their differences, have common traits. To conclude, let us return to some of those as discussed in the previous chapters. A first important point is to recognize that shamanistic knowledge is a form of thought, of philosophy, as Placide Tempels was one of the first Westerners to contend. It is based upon experiences of nature in ways that do not lead to thoughts of 'man' as the highest creature, as the one who can rule over the others, and who has a special potential for reason. On the contrary it holds fast to the inescapable dependency of human beings of the network of living forces, of vital forces, that sustain each other while eating, giving birth, passing on information, dying and transforming into food once more, etc. It understands the world, nature, as this spirited whole of life, or of love, as Swedenborg called it. To prepare for a true dialogue, Western thinkers should wane themselves from the urge to speak of spirits and of the forces of life as next to, behind or somehow 'in' visible, tangible reality. The dualisms thinkers such as Descartes and Kant left us with, should be disentangled and deconstructed.

Shamanistic knowledge has the potential to explain the world we live in, just like the modern philosophy that forms the framework of scientific knowledge. In his narrations of his contact with the Sng'oi, Robert Wolff gives a fascinating example of this, when describing the days that Ahmeed, a

Sng'oi shaman, accompanies him to the coast, as the first of his people to see the sea. He observes Ahmeed standing at the beach in silence for some time. When he returns with him to the village in the jungle, he is very surprised to hear Ahmeed, in a communal ceremony, explain all that he saw when he was there, tuning into the sea, or its spirit. Ahmeed explained to his people:

> He went on, 'The whole world is covered with the Great Ocean.' He cupped his hands about eighteen inches apart, as if to mark a globe. 'All of it covered with Ocean, and the land floats on the water.... The land is so big, there is so much land floating on this Ocean that it does not move, or maybe only a little, and we do not feel it moving.'
>
> (Wolff 2001: p. 136)

Wolff is baffled about the essentially correct view of the earth as a globe, and of the continents as being surrounded by sea, and he wonders how Ahmeed could know all of this from just staring at the sea for some time. This was not all, however, as the Sng'oi shaman continued to describe the underwater world as having underwater mountains and valleys in it, and huge streams that flow around the world. He also described the animals that live under water:

> '...There are animals so huge ... bigger than elephants.' ... 'Animals that are flat' – he clapped his hands once – 'and animals that are like snakes, but bigger, much bigger.'
>
> (Wolff 2001: p. 137)

Wolff cannot explain how Ahmeed learned from looking at the surface of the sea about its inner geography and biology. This story is important for showing that the intuitions that humans have about their world are trans-culturally available, and may not necessarily depend on empirical evidence in the modern scientific sense first. It is good to remind ourselves that the idea of the earth as a globe did not first arise in modernity, but was common to ancient peoples as well. Anthropological research has confirmed that people who live 'primitive' lives in a material sense, are not devoid of mathematical, astronomical and other knowledge considered 'scientific.'

Still there are elements in the epistemology ruling shamanistic knowledge that are not recognized in modern ideas of knowledge – most notably the idea that humans, as animated and spirited life forms, can tune in to and receive information that is passed on by other spirited life forms, be they animals, or even natural bodies such as the sea itself. For this reason I propose to speak of an anim(al)istic philosophy. It contains the word 'animistic' as used by the new animists. Their difference between new and old animists is, however, complicating things and centering or foregrounding the modern version of animism as a form of environmental ethics, foregoing the epistemological and ontological issues we have to deal with to begin to take shamanistic ways to be in the world seriously. Accepting the world as a network of life forms, and of energies of love/life, entails more than a program for moral action. In so far as it is moral, advocating certain ways to live as good, it is still based on a knowledge of how things are, with each other, connected. More elements that are recognizable throughout shamanistic ways of being in the world include the relative position of human beings – men, women, other-gendered, children, unborn, deceased. 'Man' as a universalized and essentialized demi-God, is *not* the center – instead, humans are interconnected as groups, families of generations that stretch beyond the now, including the dead as well as the unborn, which are interconnected with so many other creatures that sustain them in a fragile balance that may always turn to disaster if not cared for. Man is not the center – in order to survive humans need to understand the web of life forms from and in which they live, and help them, give to them, in order to receive what they need themselves. If the web of life is not maintained and kept, humans cannot live, let alone flourish. Man is not the center – in the human psyche/soul live many others, even though each of us has a personal identity and responsibility. These others need to be attended to, to be harmonized and cured – if necessary – to keep sane and alive and to be able to see direction. Man is not the center – humans are touching and touched creatures; they are animals that orient themselves according to inner and outer needs which, again, need to be balanced and harmonized. Animals, again, are not that different from plants, who in essence live in similar balancing acts. In fact all life does. And thus, in a sense, everything is alive.

In all that went before I refrained from giving a clear definition of the concept of nature, even though this concept is a key one in the title of this book. I could not do this differently – having written a study on the meaning of nature in modern and postmodern philosophy,[6] I was too much aware of all the different meanings it always already reflects. I needed to keep it undefined, in order to open that space for an intercultural discourse of what it means for us, culturally different groups of humans, to experience living in this world. To make clear what caring about open spaces of meaning can give, let me cite somewhat extensively what Davi Kopenawa, the Yanomami shaman, has to say about nature:

> In our very old language, what the white people call 'nature' is *urihi a*, the forest-land, but also its image, which can only be seen by the shamans and which we call *Urihinari*, the spirit of the forest. It is thanks to this image that the trees are alive. So what we call the spirit of the forest consists of the innumerable images of the trees, of the leaves that are their hair, and of the vines. It is also those of the game and the fish, the bees, the turtles, the lizards, the worms, and even the *warama aka* snails. The image of the value of growth of the forest we know as *Në*

roperi is also what the white people call 'nature.' It was created with it and gives it its richness. For us, the *xapiri* are the true owners of 'nature,' not human beings.

(Kopenawa and Albert 2013: p. 389)

In this quotation, Kopenawa does what this book intends – he enters into intercultural dialogue, into a process of translation, and of working toward shared understanding, an understanding that will necessarily contain misunderstandings and miscommunications, but this means we are communicating.

I insisted all along that an intercultural dialogue presupposes negotiations. Negotiations that focus on what we call the 'environment.' The environment is, as I hope to have shown, a contested concept, as in its secular conservationist sense, it may still transfer the colonial approach to nature and to indigenous peoples that is at issue in the first place. This is made very clear as well by Kopenawa, and again his enlightening take deserves its space here:

When they speak about the forest, these white people often use another word: they call it 'environment.' This word is also not ours and until recently we did not know it either. For us, what the white people refer to in this way is what remains of the forest and land that were hurt by their machines.... I don't like this word. The earth cannot be split apart as if the forest were just a leftover part. We are inhabitants of this forest, and if it is cut apart this way, we know that we will die with it. I would prefer the white people to talk about 'nature' or 'ecology' as a whole thing. If we defend the whole forest, it will stay alive.

(Kopenawa and Albert 2013: p. 397)

This quote makes clear that from a shamanistic perspective it makes no sense to separate ontology and ethics and focus on one of them to understand the situation we are in, as the anthropologists of the ontological turn, or the new animists, each with their specific approach, do. We would better open up to the view that, concerning our (human) relations to nature, it is never about 'conserving' something, for nature is not like a landscape captured on a photo, which tries to preserve an impression/a situation in a still-life.

The approach, the view, the understanding that is at stake in the negotiations between indigenous and moderns in their shared postcolonial situation, is that nature is a web of living relations, and that every act of any element in the web promotes its liveliness or tears it apart and renders it lifeless. Wanting to 'conserve' something, store it for 'whenever,' is already abstaining from life-promoting activities. Instead of 'conserving' humans may try to 'keep' their environment, in the meaning of keeping it up, sustaining it, feeding it, keeping it healthy and alive. To keep a forest, indigenous peoples know, humans play their role, next to the other living beings, who all together in concert maintain and promote life there. It is about planting certain trees that are loved by certain animals, that may in turn attract certain mosses and fungi that are necessary to keep the entire environment going. Planting the trees lowers temperatures, creates shade and keeps life-giving humidity around. But relating to nature does not only take place in the tropical forest. We live in many different kinds of environments. To keep a savannah-like environment, or grass-lands, wetlands or arctic icy places as areas that sustain life, we may listen to these environments that speak to those who live there and tell them how to sustain the whole. In any case, what nature tells is always local and interconnected – which is the main reason why an abstracted 'birds-eye' approach leads us away from this kind of knowledge. What shamanistic cultures and their representatives translate to modern cultures and modern peoples is basically that we, humans, are not just inhabitants but also constituents of nature – we co-constitute, keep up, communicate and move within it. This is the content of what wants to be negotiated with the moderns, with those who still think that we, humans, are living in a surrounding environment which is essentially apart from us.

Notes

1 I refer here to the title of the article by Derrida of the same name, in Derrida 1982 [1972].
2 It may be necessary for some readers to signify that this quote doesn't denounce white people as a race. It does deny their 'capitalism' however, and particularly is spoken to those who defend and wage the silent war at the fringes of modern society, to maintain the materialistic life we moderns lead.
3 Foucault, in his archeology of Kant's thought, wrote on the interlocked nature of geography, anthropology and ontology that followed from this approach:

> *On the Different Human Races* was intended to get the first lecture in physical geography of the summer semester 1775 'under way' ... But geography is not an end in itself: as an exercise, it serves as a preliminary introduction to the knowledge of the world (Weltkenntniss) that in the Anthropology Kant would later make synonymous with a knowledge of man.
>
> (Foucault 2008 [1964]: p. 32)

4 Or even a 'view from nowhere' as Thomas Nagel has called it.
5 And such a war does characterize the encounters of today's indigenous peoples with the moderns.
6 Roothaan 2005. In this book I analyzed how the concept of nature (or should we say it is a metaphor?) transfers how humans make sense of their experience of existing in a network of forces over which they do not have full control, and how this concept has been transferring many different ways to capture such experiences over the centuries.

References

Abugiche, A. S., Egute, T. O. and Cybelle, A. (2017). 'The Role of Traditional Taboos and Custom as Complementary Tools in Wildlife Conservation within Mount Cameroon National Park Buea.' *International Journal of Natural Resource Ecology and Management* 2 (3): 60–8. doi: 10.11648/j.ijnrem.20170203.13 (accessed: August 31, 2017).

Al-Assiouty, S. A. (1989). *Origines égyptiennes de Christianisme et de l'Islam*. Paris: Letouzey & Ané.

Anthony, L. and Spence, G. (2009). *The Elephant Whisperer. Learning about Life, Loyalty and Freedom from a Remarkable Herd of Elephants*. London: Sidgewick & Jackson.

Bamana, G. (2014). 'Dogs and Herders.' *Sino-Platonic Papers* 245: 1–16. Available from: http://sino-platonic.org/complete/spp245_dogs_herders_mongolia.pdf.

Barry, B. (1998 [1988]). *Senegambia and the Atlantic Slave Trade*. Cambridge, New York and Melbourne: Cambridge University Press.

Bekker, B. (1691). *De betoverde weereld*. Amsterdam: Daniel van Dalen.

Benz, E. (2002 [1948]). *Emanuel Swedenborg. Visionary Savant in the Age of Reason*. West Chester, PA: Swedenborg Studies.

Blum, D. (2006). *Ghost Hunters. William James and the Search for Scientific Proof of Life After Death*. London and New York: Penguin Books.

Boelaert, E. (1946). 'La philosophie Bantoue selon le R. P. Placide Tempels par Edmond Boelaert, avec une réaction de la part de Tempels.' *Aequatoria* 9 (3): 81–90. Available from: www.aequatoria.be/tempels/FTBoelaert.htm (accessed: October 10, 2018).

Boele van Hensbroek, P. (2001). 'Philosophies of African Renaissance in African Intellectual History.' *Quest* XV (1–2): 127–38.

Bontinck, F. (1985). *Aux origines de la Philosophie Bantoue. La correspondance Tempels-Hulstaert (1944–1948)*. Kinshasa: Faculté de Théologie Catholique. Available from: www.aequatoria.be/tempels/FTBoelaert.htm (accessed: October 10, 2018).

Bordogna, F. (2008). *William James at the Boundaries. Philosophy, Science, and the Geography of Knowledge*. Chicago and London: The University of Chicago Press.

Borg, M. J. (1994). *Meeting Jesus Again for the First Time. The Historical Jesus & the Heart of Contemporary Faith*. New York: HarperCollins Publishers.

Cabral, A. (2007 [1979]). *Unity & Struggle. Speeches and Writings*, second edition. Pretoria, RSA and Hollywood, CA: Unisa Press and Tsehai Publishers.

Craffert, Pieter F. (2008). *The Life of a Galilean Shaman: Jesus of Nazareth in Anthropological-Historical Perspective*. Cambridge, MA: James Clarke and Company.

Curry, T. J. (2017). *The Man-Not. Race, Class, Genre, and the Dilemmas of Black Manhood*. Philadelphia, Rome and Tokyo: Temple University Press.

Deacon, M. (2002 [1998]). 'The Status of Father Tempels and Ethnophilosophy in the Discourse of African Philosophy' in P. H. Coetzee and A. P. J. Roux (eds.) *The African Philosophy Reader. A Text with Readings*. London: Routledge, pp. 115–32.

Derrida, J. (1982 [1972]). 'White Mythology: Metaphor in the Text of Philosophy' in *Margins of Philosophy*. Chicago: University of Chicago Press, pp. 207–71.

Derrida, J. (2002). 'The Animal That Therefore I Am (More to Follow).' *Critical Inquiry* 28 (2): 369–418.

Derrida, J. (2006 [1993]). *Specters of Marx. The State of the Debt, the Work of Mourning and the New International*. New York and London: Routledge.

Descartes, R. (1996 [1985]). *The Philosophical Writings, Vol. I and II*. Cambridge: Cambridge University Press.

Diop, C. A. (1987). *Precolonial Black Africa*. Chicago: Lawrence Hill Books.

Dunbar-Ortiz, R. (2014). *An Indigenous Peoples' History of the United States*. Boston: Beacon Press.

Ekberzade, B. (2018). *Standing Rock. Greed, Oil and the Lakota's Struggle for Justice*. London: Zed Books.

Ekwealo, C. (2017). *Ndu Mmili Ndu Azu. An Introduction to African Environmental Ethics*. Surulere, Lagos: Redcom.

Ellis, S. and Ter Haar, G. (2004). *Worlds of Power. Religious Thought and Political Practice in Africa*. London: Hurst & Company.

Ellis, S. and Ter Haar, G. (2009). 'The Occult Does not Exist: A Response to Terence Ranger.' *Africa* 79 (3): 399–411.

Eze, E. (1997). *Race and the Enlightenment. A Reader*. Malden, MA: Blackwell Publishing.

Eze, E. (2001). *Achieving our Humanity. The Idea of a Postracial Future*. New York: Routledge.

Eze, M. O. (2017). 'Humanitatis-Eco (Eco-Humanism): An African Environmental Theory' in A. Afolayan and T. Falola *The Palgrave Handbook on African Philosophy*. New York: Palgrave Macmillan, pp. 621–32.

Fanon, F. (2004 [1961]). *The Wretched of the Earth*. New York: Grove Press.

Fanon, F. (2008 [1952]). *Black Skin, White Masks*. London: Pluto Press.

Fix, A. (1999). *Fallen Angels. Balthasar Bekker, Spirit Belief, and Confessionalism in the Seventeenth Century Dutch Republic*. Dordrecht, NL: Springer Science+Business Media.

Foucault, M. (1991 [1975]). *Discipline and Punish. The Birth of the Prison*. London and New York: Penguin Books.

Foucault, M. (2008 [1964]). *Introduction to Kant's Anthropology*. Los Angeles: Semiotext(e).

Global Witness 2015. *How Many More?* (Report, available from: www.global witness.org/en/campaigns/environmental-activists/how-many-more/ (accessed: October 18, 2018).

Goodman, F. D. (1990). *Where the Spirits Ride the Wind. Trance Journeys and Other Ecstatic Experiences*. Bloomington and Indianapolis: Indiana University Press.

Gordon, L. (2006). *Disciplinary Decadence. Living Thought in Trying Times*. Boulder, CO and London: Paradigm Publishers.

Haenen, H. (2006). *Afrikaans denken. Ontmoeting, dialoog en frictie. Een filosofisch onderzoek*. Amsterdam: Buijten & Schipperheijn.

Harvey, G. (2017 [2005]). *Animism. Respecting the Living World*. London: Hurst & Company.

Hens, L. (2006). 'Indigenous Knowledge and Biodiversity Conservation and Management in Ghana.' *Journal of Human Ecology* 20 (1): 21–30. Available from: www.krepublishers.com/02-Journals/JHE/JHE-20-0-000-000-2006-Web/JHE-20-1-000-000-2006-Abstract-PDF/JHE-20-1-021-030-2006-1561-Hens-Luc/JHE-20-1-021-030-2006-1561-Hens-Luc-Text.pdf (accessed: June 9, 2017).

James, W. (1996 [1909]). *A Pluralistic Universe. Hibbert Lectures at Manchester College on the Present Situation in Philosophy.* Lincoln, NE and London: University of Nebraska Press.

James, W. (2002 [1902]). *The Varieties of Religious Experience. A Study in Human Nature.* London and New York: Routledge.

Johnson, G. R. (2008). 'Träume eines Geistersehers – Polemik gegen die Metaphysik der Parodie der Popularphilosophie?' in F. Stengel (ed.) *Kant und Swedenborg. Zugänge zu einem umstrittenem Verhältnis.* Tübingen: Max Niemeyer Verlag, pp. 99–122.

Juhé-Beaulaton, D. (2008). 'Sacred Forests and the Global Challenge of Biodiversity Conservation: The Case of Benin and Togo.' *Journal for the Study of Religion, Nature and Culture* 2 (3): 351–72.

Juhé-Beaulaton, D. and Roussel, B. (2003). 'May Vodun Sacred Spaces Be Considered as a Natural Patrimony?' in T. Gardner and D. Moritz (eds.) *Creating and Representing Sacred Spaces.* Göttingen, Germany: Peust & Gutschmidt Verlag, pp. 33–56.

Jung, C. G. (1977). *Psychology and the Occult.* Princeton, NJ: Princeton University Press.

Kant, I. (1998 [1781]). *Critique of Pure Reason.* Cambridge: Cambridge University Press.

Kant, I. (2002 [1766]). *Dreams of a Spirit-Seer and Other Writings.* West Chester, PA: Swedenborg Foundation Publishers.

Kant, I. (2006 [1798]). *Anthropology from a Pragmatic Point of View.* Cambridge and New York: Cambridge University Press.

Kimmerle, H. (2014). 'Hegel's Eurocentric Concept of Philosophy.' *Confluence. Online Journal of World Philosophies* 1: 99–117.

Kohn, E. (2013). *How Forests Think. Toward an Anthropology beyond the Human.* Berkeley, CA and London: University of California Press.

Kopenawa, D. and Albert, B. (2013 [2010]). *The Falling Sky. Words of a Yanomami Shaman.* N. Elliott and A. Dundy (trans.). Cambridge, MA and London: The Belknap Press of Harvard University Press.

Lachman, G. (2012). *Swedenborg. An Introduction to His Life and Ideas.* New York: Jeremy P. Tarcher/Penguin.

LaLeye, I. P. (2002 [1998]). 'Is There an African Philosophy in Existence Today?' in P. H. Coetzee and A. P. J. Roux (eds.) *The African Philosophy Reader. A Text with Readings.* London: Routledge, pp. 101–14.

Latour, B. (2013). *An Inquiry into Modes of Existence. An Anthropology of the Moderns.* Cambridge, MA and London: Harvard University Press.

Lewis, I. M. (2003 [1971, 1989 2nd ed.]). *Ecstatic Religion: A Study of Shamanism and Spirit Possession.* London and New York: Routledge.

Llored, P. (2012). *Jacques Derrida. Politique et Ethique de l'Animalité.* Mons: Les Éditions Sils Maria asbl.

Magome, H. and Murombedzi, J. (2003). 'Sharing South African National Parks: Community Land and Conservation in a Democratic South Africa' in W. A.

Adams and M. Mulligan (eds.) *Decolonizing Nature. Strategies for Conservation in a Post-colonial Era*. London and Sterling, VA: Earthscan Publications Ltd., pp. 108–34.

Mall, R. A. (2000). *Intercultural Philosophy*. Lanham, MD and Oxford: Rowman & Littlefield Publishers, Inc.

Marx, K. and Engels, F. (2015 [1888]). *The Communist Manifesto*. London: Penguin Books.

Mbiti, J. S. (2006 [1969]). *African Religions and Philosophy*, revised edition. Oxford: Heinemann.

Merleau-Ponty, G. (2002 [1945]). *Phenomenology of Perception*. London and New York: Routledge.

Mosima, P. M. (2016). *Philosophic Sagacity and Intercultural Philosophy. Beyond Henry Odera Oruka*. Leiden, NL: Africa Studies Centre.

Mudimbe, V. Y. (1988). *The Invention of Africa. Gnosis, Philosophy and the Order of Knowledge*. Bloomington, IN: Indiana University Press.

Müller, L. (2013). *Religion and Chieftancy in Ghana. An Explanation of the Persistence of a Traditional Political Institution in West Africa*. Münster: Lit Verlag.

Murombedzi, J. (2003). 'Devolving the Expropriation of Nature: The "Devolution" of Wildlife Management in Southern Africa' in W. A. Adams and M. Mulligan (eds.) *Decolonizing Nature. Strategies for Conservation in a Post-colonial Era*. London and Sterling VA: Earthscan Publications Ltd., pp. 135–51.

Okwor, C. O. (2012). *The Significance of Deities in the Contemporary Nsukka Northern Igboland* (master's thesis). Available from: www.google.nl/url?sa=t&rct=j&q=&esrc=s&source=web&cd=4&cad=rja&uact=8&ved=0ahUKEwjur72KxNfOAhUJAxoKHQq9AN4QFggyMAM&url=http%3A%2F%2Frepository.unn.edu.ng%3A8080%2Fjspui%2Fbitstream%2F123456789%2F771%2F1%2FOKWOR%2C%2520CHRISTOPHER%2520OKONKWO.%2520_REV.pdf&usg=AFQjCNELn7jDfDuLzpRUBtb1LJ5YNX481A&sig2=R9t_yyUJQ9ldl3XOJdfOEQ (accessed: August 22, 2016).

Park, P. K. (2013). *Africa, Asia, and the History of Philosophy. Racism in the Formation of the Philosophical Canon, 1780–1830*. Albany: State University of New York Press.

Pedersen, M. A. (2011). *Not Quite Shamans. Spirit Worlds and Political Lives in Northern Mongolia*. Ithaca, NY and London: Cornell University Press.

Pina, I. (2001). 'The Jesuit Missions in Japan and China: Two Distinct Realities. Cultural Adaptation and the Assimilation of Natives.' *Bulletin of Portuguese/ Japanese Studies* 2 (June): 59–76. Available from: www.redalyc.org/articulo. oa?id=36100204 (accessed: December 29, 2018).

Plumwood, V. (1993). *Feminism and the Mastery of Nature*. London and New York: Routledge.

Plumwood, V. (1995). 'Human Vulnerability and the Experience of Being Prey.' *Quadrant* March: 29–34.

Plumwood, V. (2003). 'Decolonizing Relationships with Nature' in W. A. Adams and M. Mulligan (eds.) *Decolonizing Nature. Strategies for Conservation in a Post-Colonial Era*. London and Sterling, VA: Earthscan Publications Ltd., pp. 51–78.

Ranger, S. (2007). *The Word of Wisdom and the Creation of Animals in Africa*. Cambridge: James Clarke & Co.

Roothaan, A. (2005). *Terugkeer van de natuur. De betekenis van natuurervaring voor een nieuwe ethiek*. Kampen, NL: Uitgeverij Klement.

Roothaan, A. (2011). *Geesten. Uitgebannen en teruggekeerd in de moderne wereld.* Amsterdam: Uitgeverij Boom.

Roothaan, A. (2012). 'Why Religious Experience is Considered Personal and Dubitable – and What If It Were Not?' in D. Evers, M. Fuller, A. Jackelén and T. A. Smedes (eds.) *Is Religion Natural?* London and New York: T&T Clark International, pp. 117–29.

Roothaan, A. (2015). 'The "Shamanic" Travels of Jesus and Muhammad: Cross-Cultural and Transcultural Understandings of Religious Experience.' *American Journal of Theology and Philosophy* 36 (2): 140–53.

Roothaan, A. (2016). 'Political and Cultural Identity in the Global Postcolony. Postcolonial Thinkers on the Racist Enlightenment and the Struggle for Humanity.' *Acta Politologica.* Available at: http://acpo.vedeckecasopisy.cz/public Files/001247.pdf (accessed: February 2, 2017).

Roothaan, A. (2017). 'Aren't We Animals? Deconstructing or Decolonizing the Human–Animal Divide' in M. Fuller, D. Evers, A. Runehov and K.-W. Saether (eds.) *Are We Special? Human Uniqueness in Science and Theology.* New York and London: Springer, pp. 209–20.

Roothaan, A. (2018). 'Hermeneutics of Trees in an African Context: Enriching the Understanding of the Environment "for the Common Heritage of Humankind"' in J. O. Chimakonam (ed.) *African Philosophy and Environmental Conservation.* London and New York: Routledge, pp. 135–48.

Roothaan, A. (2019a). 'Decolonizing Human–Animal Relations in an African Context' in M. Chemhuru (ed.) *African Environmental Ethics: A Critical Reader.* New York and London: Springer (forthcoming).

Roothaan, A. (2019b). 'Interkulturell, transkulturell, cross-cultural – warum wir alle drei Begriffe brauchen.' *Polylog. Zeitschrift für interkulturelles Philosophieren* 40: 67–82.

Rose, D. B. (2013). 'Val Plumwood's Philosophical Animism: Attentive Interactions in the Sentient World.' *Environmental Humanities* 3: 93–109. Available from: http://environmentalhumanities.org/arch/vol.3/3.5.pdf (accessed: August 31, 2017).

Ross, E. S. (1994). 'Africa in Islam: What the Afrocentric Perspective Can Attribute to Islam.' *International Journal of Islamic and Arabic Studies* 11: 1–36.

Ross, E. S. (2006). *Sufi City. Urban Design and Archetypes in Touba.* Rochester, NY: University of Rochester Press.

Ross, E. S. (2008). 'Palaver Trees Reconsidered in the Senegalese Landscape' in M. J. Sheridan and C. Nyamweru (eds.) *African Sacred Groves. Ecological Dynamics and Social Change.* Oxford: James Currey; Athens: Ohio University Press; Pretoria: UNISA Press, pp. 133–48.

Schoonmaker Freudenberger, K. (1991). *Mbegué. The Disingenuous Destruction of a Sahelian Forest.* (Report). International Institute for Environment and Development. Paper no. 29.

Schoonmaker Freudenberger, K. and Schoonmaker Freudenberger, M. (1993). *Pastoralism in Peril: Pressures on Grazing Land in Senegal.* (Report). International Institute for Environment and Development. Drylands Program: Pastoral Land Tenure Series, no. 4.

Sheridan, M. J. (2008). 'The Dynamics of African Sacred Groves. Ecological, Social & Symbolic Processes' in M. J. Sheridan and C. Nyamweru (eds.) *African Sacred*

Groves. Ecological Dynamics and Social Change. Oxford: James Currey; Athens: Ohio University Press; Pretoria: UNISA Press, pp. 9–41.

Siebert, U. (2008). 'Are Sacred Forests in Northern Bénin "Traditional Conservation Areas?" Examples from the Bassila Region' in M. J. Sheridan and C. Nyamweru (eds.) *African Sacred Groves. Ecological Dynamics and Social Change.* Oxford: James Currey; Athens: Ohio University Press; Pretoria: UNISA Press, pp. 164–77.

Somé, M. P. (1993). *Ritual: Power, Healing and Community.* New York and London: Penguin Books.

Somé, M. P. (1994). *Of Water and the Spirit. Ritual, Magic, and Initiation in the Life of an African Shaman.* New York: Penguin Books.

Stengel, F. (2008). 'Kant – "Zwillingsbruder" Swedenborgs?' in F. Stengel (ed.) *Kant und Swedenborg. Zugänge zu einem umstrittenen Verhältnis.* Tübingen: Max Niemeyer Verlag, pp. 35–98.

Stoller, P. (1989). *Fusion of the Worlds. An Ethnography of Possession among the Songhay of Niger.* Chicago and London: University of Chicago Press.

Stoller, P. (2004). *Stranger in the Village of the Sick. A Memoir of Cancer, Sorcery, and Healing.* Boston: Beacon Press.

Storm, W. (1993). 'Bantoe-Filosofie vs. Bantu Philosophy.' *Quest* VII (2): 66–75.

Swedenborg, E. (2010 [1758]). *Heaven and Hell.* West Chester, PA: Swedenborg Studies.

Taylor, E. (1996). *William James on Consciousness beyond the Margin.* Princeton, NJ: Princeton University Press.

Taylor, C. (1989). *Sources of the Self. The Making of the Modern Identity.* Cambridge University Press.

Tempels, P. (1946). *Bantoe Filosofie.* Antwerp: De Sikkel.

Tempels, P. (1958). 'Lettre' in *Aspects de la culture noire.* Paris: Fayard, pp. 172–3. Available from: www.aequatoria.be/tempels/bio.htm (accessed: October 10, 2018).

Tempels, P. (1959 [1946]). *Bantu Philosophy.* Paris: Présence Africaine.

Tempels, P. (1962). *Notre Rencontre.* Leopoldville: Limete. Available from: www.aequatoria.be/tempels/bio.htm (accessed: October 10, 2018).

Ter Haar, G. (1992). *Spirit of Africa. The Healing Ministry of Archbishop Milingo of Zambia.* London: Hurst & Company.

Urban Walker, M. (2007). *Moral Understandings. A Feminist Study in Ethics,* second edition. New York: Oxford University Press.

Van Binsbergen, W. (2003). '"Cultures Do Not Exist." Exploding Self-Evidences in the Investigation of Interculturality' in *Intercultural Encounters. African and Anthropological Lessons towards a Philosophy of Interculturality.* Münster: Lit Verlag, pp. 459–522.

Viveiros de Castro, E. (2014 [2009]). *Cannibal Metaphysics. For a Post-Structural Anthropology.* Minneapolis, MN: Univocal.

Wolff, R. (2001). *Original Wisdom. Stories of an Ancient Way of Knowing.* Rochester, VT: Inner Traditions International.

Zhok, A. (2012). 'The Ontological Status of Essences in Husserl's Thought.' *New Yearbook for Phenomenology and Phenomenological Philosophy* XI: 99-130.

Index